BASIC HELICOPTER AERODYNAMICS

Aerospace Series List

BASIC HELICOPTER AERODYNAMICS

Third Edition

John Seddon
Formerly of the Ministry of Defence, UK

Simon Newman
Aeronautics and Astronautics, University of Southampton, UK

A John Wiley & Sons, Ltd., Publication

Registered office
John Wiley & Sons Ltd, The Atrium, Southern Gate, Chichester, West Sussex, PO19 8SQ, United Kingdom

For details of our global editorial offices, for customer services and for information about how to apply for permission to reuse the copyright material in this book please see our website at www.wiley.com.

Library of Congress Cataloging-in-Publication Data

Seddon, John M.
   Basic helicopter aerodynamics / John Seddon, Simon Newman. – 3rd ed.
       p. cm.
   Includes bibliographical references and index.
   ISBN 978-0-470-66501-5 (hardback)
  1.   Helicopters–Aerodynamics. I.   Newman, Simon, 1947- II.   Title.
   TL716.S43 2011
   629.133'352–dc22

                                                              2011010960

A catalogue record for this book is available from the British Library.

Print ISBN: 9780470665015
ePDF ISBN: 9781119994107
oBook ISBN: 9781119994114
ePub ISBN: 9781119972723
Mobi ISBN: 9781119972730

Set in 10/12pt Times by Thomson Digital, Noida, India

To Stella, for everything

Contents

About the Authors

The late **John Seddon** was a research scientist at the Royal Aircraft Establishment and then Director-General in the UK Ministry of Defence. He later became a consultant to Westland Helicopters.

Simon Newman attended Grammar School in Farnborough near the site of the Royal Aircraft Establishment. He then read mathematics at the University of Southampton, graduating in 1970. Continuing the aircraft theme, he then began a career in helicopter aerodynamics, dynamics and design for the next 41 years. Starting at Westland Helicopters, at Yeovil, Somerset, in 1970 he worked in the Aerodynamics Research Department on rotors systems, performance and aeromechanics. After a year back at Southampton in 1974, where he obtained an MSc in Aeronautics, he returned to Yeovil to work in the Aerodynamics and Dynamics Departments on rotor aerodynamics, blade behaviour and shipborne operations. He was in the Technical Office during the Falklands War, contributing to the technical backup. In 1985 he returned to Southampton as a member of academic staff, reaching the grade of Reader in 2007. His research interests have concentrated on shipborne operations, blade sailing in particular, for which he obtained his doctorate in 1995. Other research has focused on the vortex ring state and the tumble behaviour of microlight aircraft. He has several hobbies, principally photography and golf. Apart from his academic duties, he is also an Esquire Bedell of the University, carrying the mace at graduation ceremonies.

Series Preface

The field of aerospace is wide ranging and multi-disciplinary, covering a large variety of products, disciplines and domains, not merely in engineering but in many related supporting activities. These combine to enable the aerospace industry to produce exciting and technologically advanced products. The wealth of knowledge and experience that has been gained by expert practitioners in the various aerospace fields needs to be passed onto others working in the industry, including those just entering from University.

The *Aerospace Series* aims to be practical and topical series of books aimed at engineering professionals, operators, users and allied professions such as commercial and legal executives in the aerospace industry. The range of topics is intended to be wide ranging, covering design and development, manufacture, operation and support of aircraft as well as topics such as infrastructure operations and developments in research and technology. The intention is to provide a source of relevant information that will be of interest and benefit to all those people working in aerospace.

Helicopters are able to perform a wide range of roles that are not possible with conventional fixed wing aircraft, particularly due to their capability to hover, and to take-off and land vertically. There are a number of technical difficulties that have presented helicopter designers with many challenges over the years, including the aerodynamics of flexible rotors that not only provide lift, but also enable the helicopter to move forward in the desired direction.

This book, *Basic Helicopter Aerodynamics*, is the third edition of the original version that was written by the late John Seddon. Simon Newman has maintained the ethos of the first book, producing a further revision of this introductory text aimed at undergraduates and engineers new to the field that illustrates the fundamental features of rotor aerodynamics and helicopter design. Importantly, the book also maintains the balance of not delving into too much technical detail, whilst avoiding gross simplification of key important features and physical explanations. There is much to be commended in this latest expanded edition which contains a number of valuable additions to the material.

Peter Belobaba, Jonathan Cooper, Roy Langton and Allan Seabridge

Preface to First Edition

During the past decade and a half, several noteworthy textbooks have been published in the previously neglected field of helicopter aerodynamics, spurred no doubt by a growing acceptance world-wide of the importance of the helicopter in modern society. One may cite in this context Bramwell's *Helicopter Dynamics* (1976), Johnson's *Helicopter Theory* (1980) and *Rotary Wing Aerodynamics* (1984) by Stepniewski and Keys. The appearance now of another book on the subject requires some explanation, therefore. I have three specific reasons for writing it.

The first reason is one of brevity. Bramwell's book runs to 400 pages, that of Stepniewski and Keys to 600 and Johnson's extremely comprehensive treatment to over 1000. The users I have principally in mind are University or Polytechnic students taking a short course of lectures – say one year – in the subject, probably as an 'optional' or 'elective' in the final undergraduate or early post-graduate year. The object in that time is to provide them with a grounding while hopefully stimulating an interest which may carry them further in the subject at a later date. The amount of teaching material required for this purpose is only a fraction of that contained in the standard textbooks and a monograph of around 150 pages is more than sufficient to contain what is needed and hopefully may be produced at a price not beyond the individual student's pocket.

My second reason, which links with the first, concerns the type of approach. This book does not aim at a comprehensive treatment but neither is it content to consign problems to the digital computer at the earliest opportunity. In between lies an analytical route to solutions, taken far enough to produce results of usable accuracy for many practical purposes, while at the same time providing a physical understanding of the phenomena involved, which rapid recourse to the computer often fails to do. It is this route that the book attempts to follow. The analytical approach is usually terminated when it is thought to have gone far enough to serve the stated purpose, the reader being left with a reference to one of the standard textbooks in case he should wish to pursue the topic further.

The third reason is one of content. Despite the need for brevity, I have thought it worthwhile to include, in addition to treatments of the standard topics – momentum theory, blade element theory, basic performance, stability and control – a strong flavour of research and development activity (Chapter 6) and of forward-looking, if speculative, calculations (Chapter 7). It might be considered that these items are of such a transitory nature as not to be suitable for a textbook, but my criterion of stimulating the student's interest is what has determined their inclusion. Certainly they have proved to be interesting in classroom presentation and there seems no reason why that should not be so for the written word.

In addition to meeting the needs of students, to whom it is primarily addressed, the book should have an appeal as background material to short courses held in or on behalf of industry: such courses are increasing in popularity. Companies and research establishments may also find it useful for new entrants and for more established workers requiring a 'refresher' text.

Reverting to the matter of brevity, the recent publication *Helicopter Aerodynamics* by Prouty is a most admirable short exposition, well worth studying as an adjunct to any other textbook: however it shuns the mathematics completely and therefore will not suffice alone for the present purposes. Saunders' *Dynamics of Helicopter Flight* is not greatly beyond the target length but as the title implies it is concerned more with flight dynamics than with aerodynamics and is adapted more to the needs of pilots than to those of engineering students already equipped with a general aerodynamic background.

I have taken it as a starting point that my readers have a knowledge of the aerodynamics of lifting wings as they exist in fixed-wing aircraft. A helicopter rotor blade performs the same function as a lifting wing but in a very different environment; and to note the similarities on the one hand and the distinctions on the other can be a considerable fillip to the learner's interest, one which I have tried to nurture by frequent references back to fixed-wing situations. This again is a somewhat non-standard approach.

Substantial omissions from the book are not hard to find. A historical survey might have been included in Chapter 1 but was thought not necessary despite its undoubted interest. To judge by the work effort it attracts, wake analysis ('Vortex theory') deserves a more extensive treatment than it gets (Chapters 2 and 5) but here it was necessary to refrain from opening a Pandora's box of different approaches. Among topics which could have been included in Chapter 5 are autorotation in forward flight, pitch-flap coupling and blade flexibility but these were seen as marginally 'second-line' topics. The forward look in Chapter 6 might have contained a discussion of the potential of circulation control, the only system which is capable of attacking all the three non-uniformities of rotor blade flow, chordwise, spanwise and azimuthal; but the subject is too big and too distinct from the main line of treatment. The reference to autostabilization in Chapter 8 is brief in the extreme but again the choice was between this and a much lengthier exposition in which aerodynamics would have been largely submerged beneath system mechanics and electronics.

In compiling the book I have been greatly helped by discussions with Mr D.E.H. ('Dave') Balmford, Head of Advanced Engineering at Westland Helicopters, to whom my thanks are expressed. Other Westland staff members whose assistance I wish to acknowledge in specific contexts are Dr M.V. Lowson (now Professor of Aerospace Engineering at Bristol University) for Section 7.10, Mr F.J. Perry for Section 6.6, Mr R.V. Smith for Section 7.11 and Mr B. Pitkin for Chapter 8. Naturally the standard textbooks, particularly those mentioned earlier, have been invaluable in places and I trust that this fact is duly recognized in the text and diagrams.

Formal acknowledgement is made to Westland Helicopters for permission to reproduce the photographs at Figures 2.11, 4.10, 4.11, 7.6 and 7.7; to Edward Arnold, Publishers, for the use of Figures 2.10, 2.13, 5.1, 5.3, 6.3, 8.5 and 8.6 from A.R.S. Bramwell's book *Helicopter Dynamics* (1976); to Mr P.G. Wilby of the Royal Aircraft Establishment for Figures 6.2 and 6.5, which are reproduced with the permission of the Controller of Her Majesty's Stationery Office; and to Dr J.P. Jones for the use of Figures 2.12, 4.2 and 4.4.

My thanks are due to Molly Gibbs of Bristol University who copy-typed the manuscript and to my grandson Daniel Cowley who drew the figures.

                                                                                        J. Seddon

Preface to Second Edition

The original *Basic Helicopter Aerodynamics* was conceived and written by Dr John Seddon. It found a respected place in the subject of rotary wing aircraft and has informed many. Sadly John Seddon has since passed away and I was very flattered to be asked to revise his manuscript for a second edition. This brought an immediate problem. Do I strip the work down to nuts and bolts or do I revise it as it stands but add my own contributions? Since the book is now under joint authorship, it would have been unfeeling to have pursued the former option since the original concept of John Seddon would have disappeared. For that reason I decided to pursue the latter option of revising the text and adding to it – particularly in the field of illustrations. The design, manufacture and operation of the helicopter rotor tend to be rather esoteric for the newcomer and long textual descriptions can be dry and not helpful. I have added, therefore, a substantial number of images to illustrate and clarify the discussions.

The original diagrams were created by hand, which did not altogether succeed. Since that time, computer technology has improved greatly and the book's graphics have been updated accordingly. The book's size has increased to allow for the additions but I have been mindful of the need to retain the compactness of the original work.

Helicopter rotor aerodynamics continues to be investigated. It is essential to introduce recent developments to the student and I intend to maintain this book in a form that will introduce the latest developments. While an introductory text cannot hope to describe new techniques in detail it must be capable of establishing the correct thoughts in the reader's mind, thus preparing them for more intensive study.

The revisions have been aimed at illustrating, more fully, the various features of rotor aerodynamics and helicopter design. The helicopter is unique in its linking of the aerodynamic and mechanical features and a full appreciation of these air vehicles can only be achieved by understanding these interactions. Many of the extra figures illustrate the diversity in the design and operation of a helicopter and these differences are highlighted in the text.

As with all things aeronautical, a team effort is always needed, and the assembly of this book is no exception. A picture says a thousand words so I have called upon the skills of many people to provide as many photographs as possible to amplify and, hopefully, clarify the explanations. While I have been able to supply a number of these photographs personally, a considerable number have been kindly supplied and I would like to sincerely thank the following people for their generosity. Denny Lombard of Lockheed Martin, Alan Vincent, Alan Brocklehurst and Alan Jeffrey of GKN Westland Helicopters, Harry Parkinson of Advanced Technologies Incorporated, Stewart Penney, Guy Gratton, David Long of Kaman and Steve Shrimpton.

While I am quite pleased with my own photographic attempts, I am mindful that the pictures were taken on the ground, usually on a pleasant warm day with plenty of time to press the shutter release. In contrast, the above mentioned people have obtained better quality results while often hanging out of an aircraft in very difficult situations. This marks the difference between the amateur and the true professional.

I would also like to thank my colleagues and researchers who have provided much thought provoking discussion, which I hope, is reflected in the book. I am very grateful to David Balmford for his suggestions in correcting the text. I also would like to express my thanks to Ian Simons for his constant advice on all matters aeronautical. I offer many thanks to Julia Burden at Blackwell Science for her forbearance. The manuscript was late and she stuck with it, probably biting her lip but giving me valuable support. She offered me the task of revising the book and I hope she is not disappointed.

Finally I would like to thank my wife, Stella, for putting up with my constant whizzing around putting the final touches to this work, snatching a cup of coffee as I speed by.

Simon Newman
Winchester
January 2001

Preface to Third Edition

The first edition of *Basic Helicopter Aerodynamics* was written by John Seddon and quickly found a place in educating new helicopter engineers and technologists in addition to undergraduates and postgraduates. Very sadly his early demise prevented him from seeing his creation develop. The publishers kindly approached me to conduct the first revision of the book and it was completed nearly a decade ago. It was a real pleasure to provide my own input to the book's evolution; it was certainly daunting but I believe it still has a firm place in the helicopter world. This same daunting feeling returned about two years ago when the present publishers asked me to prepare a third edition. It is imperative in an introductory text, as this book is, not to take the reader too far into the fine details of the subject; however, it is being unfair to lightly touch on the subject and gloss over important factors which link the various theories and analyses together.

The linking of technical methods is particularly relevant in the helicopter since it is not possible to isolate the aerodynamics of the rotor and overall aircraft from the dynamic responses of the blades themselves. Each influences the other and the complete problem has to be solved.

In 1970, after graduating in mathematics, I arrived at Westland Helicopters in Yeovil to begin my career. I walked into a subject in which I had the mathematical skills but very soon became aware that I needed to learn how to apply that mathematical knowledge. I also soon realized that I was working with some very sharp minds and, with their help, encouragement and a wealth of experience became a colleague. It is due to their dedication and generosity that I came to build a career in helicopter aeromechanics – for which I will always be grateful. I was well schooled in the intricacies of helicopter rotors, initially by Geoff Byham, Ian Simons and Bob Hansford. As my career developed I enjoyed the company of colleagues Alan Vincent, Steve King, Tom Beddoes and John Perry.

I have had many interactions with other academics, researchers, those in the armed forces and a whole host of flyers, and they too have my gratitude. In academia I learnt a great deal from Geoffrey Lilley, Ian Cheeseman, Roddy Galbraith, Roy Bradley, Gordon Leishman, Gareth Padfield and Richard Brown. Working in the helicopter industry allows contact with experts in their field, such as the late Peter Wilby of the RAE, Tim Cansdale of dstl and David Lee of the Empire Test Pilots' School. One of the great pleasures of working in a university is seeing young minds develop and I was lucky enough to see Ajay Modha, Malcolm Wallace, Mark Jones, Peter Knight and Matthew Orchard through their earliest days in helicopter research and into the aerospace industry. I sometimes felt a twinge of envy when, perhaps with another start, I could have made a greater contribution. However, they are the young minds that will define the future and I will leave it in their capable hands.

To create a book, you need help and contributions and I would like to detail them now. If we start with the text, then the staff of the university must be acknowledged. I am very grateful to my Head of School, Mark Spearing, for encouraging me to undertake this task. I am in debt to my colleagues, Scott Walker and Hazel Paul, who have been instrumental in keeping me on the rails. Their contribution was to get a set of chapters assembled, in a sensible time, and which read well. This is vital for an introductory text. I owe them some favours. I was very fortunate to have David Lee of the Empire Test Pilot's School check Chapter 8 and make valuable suggestions to help its readability. The book contains many images and while I was able to produce some myself, I needed to ask for the generosity of many skilful photographers to fill in the many gaps. David Gibbings of Agusta Westland has been very generous with his time in providing images and technical support. In digging around on the Web I encountered high-quality photographic work from Ashai Bagai, Steve Rod, Markus Herzig, John Olafson and Stewart Penney who kindly let me make use of their hard work. John Piasecki of the Piasecki Corporation was very supportive by providing two images of its compound helicopter designs. Christina Gotzhein of Eurocopter was very helpful in supplying two images of its helicopters either in the Himalayas or with their latest design – only just past its first flight! Paul Oelkrug of the McDermott Library at the University of Texas provided the image of the XC 142 and I am grateful for all of his efforts.

Finally, I would like to express my sincere appreciation of the US Navy and Air Force websites which contain a gallery of really outstanding images and which are placed in the public domain. They have made a significant contribution to the content of this book.

I would like to thank Debbie Cox and Eric Willner of John Wiley & Sons, Ltd for their encouragement and willingness to grease the works, which relieves me of many factors required to get a book onto the bookshelves.

I would also like to record my gratitude to one of their colleagues, Nicky Skinner. Nicky provided my main interface with John Wiley & Sons, Ltd and helped me enormously with the nuts and bolts that authors all too easily forget. She also lit the odd fire under me when my mind drifted onto other things and deadlines were approaching at an alarming rate. I was looking forward to seeing it through with her and only recently we met to go for the final push to the book hitting the shelves. She was young, bright and delightful and knew her job inside out, so it was with great sadness I learned that she had passed away, suddenly and all too early. In a fairer world she had her life before her and many authors to encourage and help over the finish line. It is a great shame that this can no longer happen and I would like to acknowledge the many, and valued, contributions she made to the genesis of this book. I will miss her charm and above all her smile. The task of getting the book from computer to the page requires a considerable effort from the production staff and I am very grateful to Genna Manaog for helping the book through its final run in to the printing press. The difference between what you want to say and what you actually say is always a problem. It takes a good copy editor to straighten out the words and I am indebted to Neville Hankins for his contributions.

I have one final acknowledgement to make. My wife Stella has provided me with considerable support over my working life but due to a severe illness has not been able to take part in putting this book together in the normal way. However, I have had her spirit with me throughout and I would hope that she would, under normal circumstances, have approved of the final result. I would like to dedicate this book to her.

In looking at the final version of this book, I am reminded of the long road I have travelled. I started working on helicopters because, quite honestly, I had but one job offer after graduation and that was in the Aerodynamics Research Department at Westland Helicopters. So on 2 September 1970, wearing a brand new suit from Dunn and Co., and on a sunny morning in Somerset, I plunged headlong into the subject. So writing this on 18 February 2011, there are 41 years gone by; sometimes it feels like 141 years and sometimes like 41 days. In either case it has been interesting, difficult, frustrating, maddening but ultimately worthwhile, enlightening, humbling and fun. I have said earlier that I have had support from many colleagues and I consider that I have been lucky indeed. I hope there are still more hurdles to clamber over and while there is a neuron or two left I will keep at it. To you the reader, there is still much to do and I wish you well. I sincerely hope that, in reading this book, you will be encouraged and stimulated. It will never be easy but then, if it was, there would be no satisfaction. If you keep at it there will be the occasional Eureka moment and those are the times when you will feel fulfilled.

Bon voyage.

Simon Newman
Winchester
March 2011

Notation

General

a	lift curve slope $dC_L/d\alpha$
a_0	first term in Fourier expansion of β
a_1	coefficient of second term in Fourier expansion of β
a_2	coefficient of fourth term in Fourier expansion of β
A	area of rotor disc
A_b	total blade area (N blades)
A_1	coefficient of second term in Fourier expansion of θ
A_2	coefficient of fourth term in Fourier expansion of θ
A_p	projected frontal area of rotor head (Chapter 6)
A_s	flow spoiling factor (Chapter 6)
A_z	boundary layer shielding factor (Chapter 6)
b_1	coefficient of third term in Fourier expansion of β
b_2	coefficient of fifth term in Fourier expansion of β
B	tip loss factor in $r = BR$
B_1	coefficient of third term in Fourier expansion of θ
B_2	coefficient of fifth term in Fourier expansion of θ
c	blade chord
C_D	drag coefficient
C_L	lift coefficient
C_H	H-force coefficient
C_P	power coefficient
C_Q	torque coefficient
C_T	thrust coefficient
d	differential operator
D	aerodynamic drag
e	hinge offset ratio
f	equivalent flat-plate area
H	H-force
I	moment of inertia
k	empirical constant in expression for profile power
K	empirical constant in Glauert expression for induced velocity
l	moment arm of tail rotor thrust about main shaft
L	aerodynamic lift
m	blade mass per unit span
M	figure of merit
M	Mach number

M	moment (Figures 8.4 and 8.5)
M_{T}	aerodynamic moment about flapping axis
n	inertia number (Chapter 8)
N	number of blades
p	static pressure
P	power
q	dynamic pressure, $\frac{1}{2}\rho V^2$
q	torque coefficient (Bramwell definition)
Q	torque
R	blade radius
S	stiffness number
t_{c}	thrust coefficient (Bramwell definition)
T	thrust
u	component velocity (non-dimensional, $U/\Omega R$)
U	component velocity (dimensional)
v	induced velocity
V'	hypothetical velocity in Glauert formula for forward flight
V_{c}	climbing speed
V_{i}	stream velocity (flight speed)
w	disc loading, T/A
W	aircraft weight
x	fraction of blade span from axis $(=r/R)$
y	distance along blade span from axis
z	height of rotor plane above ground

Greek

α	incidence (angle of attack) of blade, positive nose-up
α	incidence of fuselage (Chapter 6), positive nose-up
α_{r}	angle of attack of tip path plane to flight direction, positive nose-down
β	compressibility factor (Chapter 7)
β	flapping angle (blade span to reference plane)
γ	Lock number
δ	relative density of air, ρ/ρ_0
Δ	prefix denoting increment, thus ΔP
ζ	lag angle
θ	blade pitch angle
κ	empirical constant in expression for induced power
λ	blade natural flapping frequency (Chapters 4 & 8)
λ	inflow factor (non-dimensional induced velocity)
μ	advance ratio, $V/\Omega R$
π	pi
ρ	absolute density of air
σ	blade solidity factor
ϕ	angle of resultant velocity at blade to reference plane
χ	sweep angle
ψ	angle of azimuth in blade rotation
Ω	blade rotational speed, radians per second

Suffixes

av	available
b	blade
c	suffix for thrust coefficient (Bramwell definition)
C	in climb
D	drag
D	descent
h	hover value
H	H-force
i	induced
L	lift
Max	maximum
o	basic or constant value
p	parasite
P	power
Q	torque
req	required
t	blade tip
tw	blade twist
T	thrust
∞	conditions 'at infinity', that is where flow is undisturbed
0	hover value

Units

The metric system is taken as fundamental, this being the educational basis in the UK. Imperial units are still used extensively, however, particularly in the USA but also by industry and other organizations in the UK. For dimensional examples in the text and diagrams, therefore, those units are used which it is felt have stood the test of time and may well continue to do so. Often units in both systems are quoted; in other cases reference may need to be made to the conversion tables set out below. In either system, units other than the basic one are sometimes used, depending on the context; this is particularly so for velocity, where for example aircraft flight speed is more conveniently expressed in kilometres/hour or in knots than in metres/second or in feet/second. The varieties used in the book are included in the table.

Quantity	Metric unit and symbol	Imperial equivalent
Primary quantities:		
Mass	kilogram (kg)	0.0685 slug
Weight	newton (N)	0.2248 pound (lb)
Length	metre (m)	3.281 feet (ft)
Time	second (s)	1.0 s (sec)
Temperature	kelvin (K)	Celsius ($^\circ$C)
Temp(K) = temp ($^\circ$C) + 273.15		
Derived quantities:		
Weight (force)	kilogram force	
	9.807 N (kg)	2.2046 lb
Density	kg/m^3	0.00194 slug/ft^3
Pressure	N/m^2	0.0209 lb/ft^2
	0.1020 kg/m^2	
Velocity	m/s	3.281 ft/sec
	3.600 km/h	196.86 ft/min
		1.941 knots
Acceleration	m/s^2	3.281 ft/sec^2
Accel. of gravity	9.807 m/s^2 (g)	32.2 ft/sec^2
Power	watt, N m/s (W)	0.7376 ft lb/sec
Metric horsepower	75 kg m/s (mhp)	0.9863 HP
English horsepower	76.04 kg m/s	550 ft lb/sec

Abbreviations

ABC	Advancing Blade Concept
ADT	Actuator disc theory
Aero. J.	*Journal of the Royal Aeronautical Society*
AFCS	Automatic flight control system
AGARD	Advisory Group for Aerospace Research and Development
AIAA	American Institute of Aeronautics and Astronautics
ARC	Aeronautical Research Council
ASE	Automatic stabilization equipment
AUW	All-up weight
BERP	British Experimental Rotor Programme[1]
BET	Blade element theory
CFD	Computational fluid dynamics
CG	Centre of gravity
DOF	Degree of freedom
FAI	Fédération Aéronautique Internationale
FOM	Figure of merit
FUMS	Fatigue and Usage Monitoring System
HUMS	Health and Usage Monitoring System
IFR	Instrument Flight Rules
IGE	In ground effect
ISA	International standard atmosphere
JAHS	*Journal of the American Helicopter Society*
NACA	National Advisory Committee for Aeronautics (now NASA)
NFA	No-feathering axis
NFP	No-feathering plane
NPL	National Physical Laboratory
OGE	Out of ground effect
PIV	particle image velocimetry
RAE	Royal Aircraft Establishment
RAF	Royal Air Force

---

[1] This was a collaborative programme, jointly sponsored by Westland Helicopters and the Ministry of Defence, which embraced blade aerofoil section research between the Royal Aircraft Establishment (now Dstl) and Westland Helicopters. This research work developed the advanced rotor blade and tip design for the Westland Lynx helicopter. Its most public achievement was the world speed record of 249.1 mph by aircraft serial number XZ170, designation G-LYNX, on 11 August 1986 over the Somerset Levels.

RLD	Really low drag
R & M	Reports and Memoranda of the ARC
SAE	Society of Automotive Engineers
SA	Shaft axis
sfc	Specific fuel consumption
SHM	Simple harmonic motion
SL	Sea level
SNP	Shaft normal plane
TPA	Tip-path axis
TPP	Tip-path plane
TRLF	Transmission loss factor
UAV	Unmanned aerial vehicle
UK	United Kingdom
USA	United States of America
VRS	Vortex ring state
VTOL	Vertical take-off and landing
WHL	Westland Helicopters Ltd

1

Introduction

1.1 Looking Back

1.1.1 Early Years

The first foray into rotary-winged flight occurred around 400 BC with a toy known as the Chinese top. It was constructed with a central shaft to which was attached wings consisting of feathers or flat blades inclined to the rotation plane normal to the shaft. This was spun between the hands and by generating thrust flew for a short period of time. Jumping forwards to AD 1325, a Flemish manuscript contained the first known illustration of a helicopter rotor which was operated by the pulling of a string.

Just over 150 years later we encounter probably one of the most influential milestones in the work of Leonardo da Vinci. His famous design of a rotary-wing vehicle, see Figure 1.1, forms an ideal illustration of the origin of the term 'helicopter'. It is commonly considered to be the combination of the words 'helix' and 'pteron' giving the concept of the 'helical wing'.

In 1784, Launoy and Bienvenu built a mechanical model with two rotors. It was effectively the Chinese top but extended to a pair of rotors on the same axle but rotating in opposite directions. It was powered by a leaf spring and strings. It is similar to the coaxial configuration of more modern times.

In 1810, Sir George Cayley wrote an aeronautical paper preparing the ground for future helicopter development. He designed an air vehicle consisting of two pairs of contra-rotating rotors on either side of a canvas-covered central fuselage structure to generate lifting thrust. The rotor blades were inclined sectors with no noticeable aerofoil. Forward propulsion was provided by a pair of pusher propellers mounted at the rear. This design was the forerunner of many features of modern rotary-winged vehicles. The absence of a sufficiently powerful engine prevented the project from leaving the drawing board – let alone the ground.

The year 1878 saw Forlanini build a model powered by steam. It flew to a height of 40 feet for a period of 20 seconds. It was impossible to achieve this by carrying a steam engine – because of the weight – so Forlanini devised a method whereby a sphere was positioned underneath the model and provided a storage vessel for the steam. This was then slowly tapped to obtain the required propulsive torque.

The lack of suitable power also hampered the work of Thomas Alva Edison. He initially conducted tests on rotors to examine the thrust v power performance. Finding this to be unfavourable he then focused on the engine and wrote of his attempts to use gun cotton in the

Basic Helicopter Aerodynamics, Third Edition. John Seddon and Simon Newman.
© 2011 John Wiley & Sons, Ltd. Published 2011 by John Wiley & Sons, Ltd.

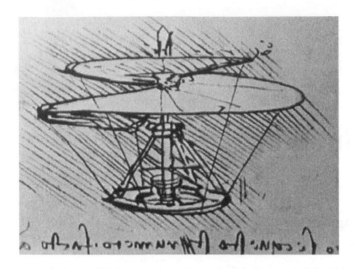

Figure 1.1   Leonardo da Vinci's helicopter concept

cylinder of an engine fired with a spark. He obtained good results but injured himself and one of his colleagues by singeing! Edison was not daunted and later work provided estimates of the required power/weight ratio that would allow a flyable helicopter to be achieved.

The absence of a powerful enough engine dogged the early 1900s; however, in September 1907 Louis Breguet achieved a milestone of the first man lifted with a tethered rotary-winged aircraft. The altitude was a mere 2 ft, but a 1 minute flight was the result. However, stability was obtained by four assistants actually holding the aircraft. He later built two other air vehicles but suffered from lack of sufficient power and lack of control both in flight and in landing.

November 1907 saw Paul Cornu achieving what is often considered to be the first 'true flight' of a helicopter. The vehicle rose to an altitude varying between 1 and 5 feet for a period of 20 seconds. It was fitted with two rotors of approximately 20 feet diameter mounted in a tandem layout – see Figure 1.2. Forward propulsion was provided by vanes positioned underneath the

Figure 1.2   Cornu helicopter

rotor discs which deflected the rotor downwash backwards and downwards. The efforts of Breguet and Cornu highlighted the importance of stability and control in flight.

In 1909, a pioneering figure appeared who was to become very influential in the history of the helicopter. He was Igor Sikorsky, who initially built two helicopters which proved unsuccessful. This was the short-term result, but it taught the Russian about what to do and what not to do. In the interim period he moved to fixed-wing aircraft but 20 years later he was to bring the lessons of the past into focus and become a dominant figure in helicopter technology.

The year of 1912 saw the Danish aviator Ellehammer use cyclic pitch, successfully giving stable and controllable flight for the first time. The vehicle was lifted by two contra-rotating rotors on the same axle. The construction of both was of a ring to which was fitted the blades. The lower rotor was fitted with fabric to increase the lift force. Forward flight was aided by a conventional propeller.

1.1.2   First World War Era

In the First World War, Petroczy built a vertically lifting machine hoping to replace captive observation balloons, which were very vulnerable. The technical and experimental development was conducted by the illustrious Theodore von Karman. It consisted of two contra-rotating rotors positioned within a framework which also contained three radial engines and the undercarriage was pressurized bags placed in the centre and at the end of three legs. An observation basket was placed above the rotors. Wilhelm Zurovec is often neglected in any description of this air vehicle except that the designation is PKZ (Petroczy Karman Zurovec).

In May 1918, the PKZ2 (see Figure 1.3), Hungary's first military helicopter, flew tethered to a height of 50 m with new 31 hp Le Rhone engines.

1.1.3   Inter-war Years

In 1921 George de Bothezat directed the first US Army programme into helicopters. After working in secrecy at Dayton, the de Bothezat helicopter (see Figure 1.4), made its first flight in December 1922.

Figure 1.3   PKZ 2

Figure 1.4   The de Bothezat Quadrotor

The rotor layout was to have one rotor on each of four arms with the pilot positioned in the centre. Numerous flights were made in 1923 carrying up to four people. The US Army did not pursue the design but acknowledged its contribution to helicopter technology generally.

In the 1920s Emile Berliner developed a helicopter with the usual, at that time, layout of contra-rotating rotors with vertical and horizontal vanes positioned under the rotor aiming, unsuccessfully, to give control. In 1922, the design moved into placing rotors on wing tips fitted to what was essentially a fixed-wing fuselage based on a Nieuport 23. Pitch control was achieved by a small propeller located just forward of the fin. The problems faced were those of not sufficient rotor size to generate sufficient thrust and blockage of the rotor downwash by the wing surfaces.

The French engineer Etienne Oehmichen began helicopter experiments by using balloons to assist the rotors in lifting the aircraft. In 1922, he was able to discard the balloon and use a four-rotor configuration. There were also five small, horizontal, variable pitch propellers to control the aircraft's attitude in flight. Many flights were made lasting several minutes and in April 1923 the record distance of 358 m was achieved. In April 1924 Oehmichen raised the straight line distance record to 525 m only to have it broken the following day by the Spaniard, Marquis Pateras Pescara. In the following May he succeeded in flying a 1 km closed circuit course winning a 90 000 franc prize from the Service Technique de L'Aeronautique.

Pescara flew for 736 m with a design of four biplane rotor blades on each of two rotors. These were placed on the same vertical axle and rotated in opposite directions. The pilot had control over the blade pitch and this helicopter was important in this respect as it heralded the use of autorotation for safe landing in the event of engine failure. Pescara attempted the closed course flight but was unsuccessful and it was left to Oehmichen to achieve this the following year.

In England Louis Brennan was developing his helicopter at the Royal Aircraft Establishment in Farnborough – see Figure 1.5. It was a two-bladed rotor powered by propellers fitted to the blade tips.

Thus external propulsion did not need torque reaction of a second rotor. Tethered flights inside a hangar were undertaken in 1924 with free flights beginning in 1925. They were halted in 1926 after an accident.

In 1925, the Dutch engineer Baumhauer developed a single two-bladed main rotor helicopter which had the beginnings of blade control via a swash plate system. It also had torque control provided by a separate vertical rotor, which was the forerunner of the modern day tail rotor. The flight of the helicopter was not altogether successful and chains were hung from the corners of the airframe to improve the stability. While ultimately unsuccessful, this design provided important technological pointers for the future.

The Hungarian engineer Oscar von Asboth developed a sequence of designs which comprised two contra-rotating main rotors and control was achieved using vanes placed in

Figure 1.5 Brennan helicopter

the downwash. The fourth variant was the AH4 and in 1930 it climbed to a height of 31 m, flying a distance of 2800 m. Also in 1930 the Italian Corridon D'Ascanio broke three world records of altitude (59 feet) distance (1180 yards) and endurance (8 minutes, 45 seconds). The configuration consisted of two contra-rotating main rotors of two blades each. The blade pitch was controllable in flight by the pilot by means of trailing-edge-mounted tail elevators – a unique design feature at that time.

Also in the 1930s the Belgian Nicolas Florine was working on what is now termed the 'tandem helicopter layout'. Both rotors spun in the same direction and the counteracting torque was provided by tilting the two rotor shafts appropriately.

Breguet now reappeared working with Rene Dorand. The Breguet–Dorand Gyroplane was so successful in its flight that many feel that it is the first real helicopter. It made its first flight in June 1935 after which flight tests and design amendments were made over the following months. December 1935 saw this helicopter presented to the world and before the FAI it flew at 67 mph raising the world speed record further. In September 1936 it rose to a record 517 feet altitude and in November set the closed course record of 27.4 miles. The design had two twin-bladed rotors which were mounted on universal joints allowing the rotor discs to be used to control the helicopter flight. This is the beginning of the articulated rotor design.[1] Torque control was achieved by differential blade pitch on the two main rotors.

---

[1] Helicopter control is based primarily on directing the thrust generated by the main rotor(s). It is normally accepted that this thrust force is in a direction perpendicular to the rotor disc – that is, the plane traced out by the blade tips. This plane can be altered in two ways. The first is to alter the direction of the rotor shaft taking the rotating blades with it. The second is to permit the blades to move in a direction normal to the rotation plane, known as flapping, which allows the rotor disc plane to change without requiring any direct change in the rotor shaft. The tilting of the thrust force will in due course cause the helicopter to change its flight attitude which will then take the rotor shaft with it.

The final part of the early years of helicopter development is defined by the work of Henrich Focke in Germany. One year after the first flight of the Gyroplane, the Focke–Achgelis Fa61 flew for the first time. This used two three-bladed main rotors placed laterally on outriggers – the forerunner of the side-by-side configuration. It was very successful in breaking world records including 77 mph, 8000 feet altitude and an endurance of 1 h 20 min. It also flew a closed course of 50 miles. In February 1938, Hanna Reich flew this aircraft inside the Berlin Deutschlane Halle sports arena. It took off, hovered and flew sideways within the hall in front of many spectators. With the banked seating, recirculation was very apparent after a period of flight, with the consequence that much rubbish was seen flying through the air!

Alongside the development of the helicopter, the autogyro played a significant part in the evolution of its sibling. In January 1923 the Spaniard Juan de la Cierva achieved the first flight of an autogyro. The vehicle itself was developed from the airframe of a fixed-wing aircraft; the wings were shortened, a rotor was fitted above the centre of the fuselage and the empennage was retained virtually unmodified. It was designated the Cierva C-4. de la Cierva is one of the most respected rotorcraft pioneers and his work was very influential in the history of rotorcraft engineering. In September 1928 the first rotary-winged aircraft crossing of the English Channel took place when de la Cierva flew his C8L MkII, including a passenger from Croydon, to Le Bourguet. Figure 1.6 shows a later type – the C8V.

In the Soviet Union, the first tethered helicopter test took place in August 1930 with the Ts AGI1-EA.

In January 1939 a Kellett autogyro flew the first mail service for Eastern Airlines linking the main Post Office in Philadelphia, PA with Camden NJ Airport.

1.1.4   Second World War Era

In December 1941 Igor Sikorsky flew his VS 300 helicopter design – see Figure 1.7 to a world endurance record of over 1 hour. It was fitted with collective and cyclic pitch on the main rotor and had a system of rotors at the tail. The torque reaction was provided by a vertical tail rotor and a horizontally aligned second rotor provided adjustment in pitch. In January 1942 an improved VS 300 flew for the first time. Designated the XR4 it was the forerunner of the R4 which was the first production helicopter outside of Germany. It was also the only

Figure 1.6   Cierva C8V autogyro (Courtesy Solent Sky)

Figure 1.7   Vought Sikorsky VS300

helicopter operated by the USA during the Second World War and became known as the Hoverfly in the UK.

In 1943 the Doblhoff WNF 342 emerged. It used tip-jet drive with fuel being pumped from tanks in the fuselage along hollow rotor blades and burnt at the tips.

In October 1943 the R4 (designated HNX-1) became the first helicopter to be accepted by the US Navy.

In 1944 the Flettner rotor appeared on the Kolibri FL282 and the German Ministry of Defence placed an order for 1000 aircraft. This type of twin rotor is seen today in the Kaman Huskie and KMax helicopters and consists of two identical rotors, rotating in opposite senses. They have rotor shafts placed close together – laterally – and inclined outwards enabling the rotors to turn without any blade clashing.

January 1944 recorded the first mercy mission when a R4 US Coast Guard helicopter flew blood plasma to aid 100 crew members injured in an explosion on a US Navy destroyer.

September 1945 saw the first helicopter flight across the English Channel when a German Crew flew a Fa 223 Druck (Dragon) to Airborne Force Experimental Establishment. This was a large utility helicopter with two rotors placed on outriggers.

During the period of the Second World War, Sikorsky produced 600 airframes of their XR4, XR5 and XR6A variants before VJ Day on 2 September 1945.

1.1.5   Post-war Years

This period saw the emergence of a number of helicopter engineers whose names remain in the forefront of rotary-wing technology today. Bell had Arthur Young, who developed the Bell Bar system of rotor control. Frank Piasecki provided advancement of the tandem configuration with his own company where the compound configuration was added to the portfolio. Charles Kaman, also establishing his own helicopter company, took the Flettner type of rotor forward and also developed conventional rotorcraft layouts. His aircraft designs differed from the mainstream in that the rotor blade pitch control was achieved by using servo flaps placed on the trailing edge of the blades. This caused blade pitch change to be achieved via elastic twisting rather than using a pitch bearing at the rotor head itself. Another helicopter name was borne by

Figure 1.8   Bell 47 (Courtesy John Olafson)

Stanley Hiller. His helicopters used a method of rotor control where a servo paddle was placed underneath the rotor, in an orientation similar to the Bell Bar of Arthur Young. In the Soviet Union we have the names of Mil, Kamov, Yakovlev, Archangleis and Tischenko appearing.

The conditions under which helicopters could operate were extended by the development of de-icing to main and tail rotors, engine inlets, the use of radar altimeters, long-range navigation and Instrument Flight Rules (IFR). The loads carried by helicopter components were becoming heavier and so the move from forgings to castings was being made for rotor heads and transmission.

On 8 December 1945 the Bell 47 – see Figure 1.8 – made its first flight beginning a long contribution to helicopter development. In May 1946 we find the appearance of the first US commercial helicopter in the guise of the Bell 47 CAA Type Cert H-1.

In 1946 W. Laurence LePlage produced the XR1 side-by-side rotorcraft, which was notable since it extended to a tilt rotor configuration. Robert Lichten with the Guerierri Transcendental aircraft model 1G appeared the previous year and he eventually moved to Bell to take responsibility for tilt rotor development.

The year 1947 saw the first all-metal helicopter – the Kellett XR-10 – which became the first twin-engine rotary-winged aircraft for the US Air Force.

Later that year there was the appearance of the McDonnell XH-20 Little Henry – see Figure 1.9. This was the first tip-jet-powered helicopter to be driven by ramjets. This had the disadvantages of a high fuel consumption and noise production, a problem with tip-jet drives which surfaces later in the history of the helicopter.

The following year, in August, saw the Sikorsky S52 become the first helicopter to use all-metal rotor blades. The following month, the Mi-1 (Hare) became the first mass-produced Soviet helicopter.

In 1949, a Hiller Model 12 flew the first transcontinental flight across the USA by a commercial helicopter. Also that year, in the UK, the first commercial helicopter began operation after the Bristol Type 171 Sycamore gained a commercial certificate of airworthiness.

In Europe, it was considered that Second World War designs could not be developed satisfactorily, so in 1945 England and France promoted new helicopter concepts. In 1950 Fairey considered the Doblhoff concept of the tip jet. It produced the Fairey Ultralight which

Figure 1.9 McDonnell 'Little Henry'

was successfully demonstrated landing and taking off from a lorry at the Farnborough Air Show. This design was stopped in favour of the W14 Skeeter being produced by the Cierva Autogyro Company in Eastleigh/Southampton in England. The Cierva Company had produced its W9 helicopter – see Figure 1.10 – in 1946, which had a novel feature in the tail design. Rather than a conventional tail rotor for torque reaction, it used the tail boom as a plenum chamber and expelled air via a sideways-facing jet. This concept is now used today as the NOTAR system. It also used a different concept for main rotor control. The blades were attached to the hub with a fixed pitch angle which places thrust control on the rotor speed and has the inherent problem of a high rotor inertia preventing rapid changes in rotor speed and hence rotor thrust. The main rotor disc attitude was controlled by physically orientating the main rotor hub into the rotation

Figure 1.10 Cierva W9 (Courtesy Solent Sky)

Figure 1.11   W11 Air Horse (Courtesy Solent Sky)

plane required – as already described. This required a significant effort from the pilot and powered controls were needed.

The W11 Air Horse was revolutionary in that it had three main rotors. A rather unusual layout, it was born of a combination of the tandem and side-by-side rotor layouts. It flew from Southampton but ultimately suffered a crash near Romsey about 10 miles from Southampton in June 1950. Figure 1.11 shows the Air Horse.

Fairey Aviation Co. produced a sequence of novel helicopter designs starting in 1947 with the Gyrodyne. This belonged to a type of helicopter known as a compound. In these designs, the main rotor forward propulsion is augmented by a separate propulsive device. In the Gyrodyne, the main rotor torque reaction was achieved with a forward-facing propeller placed on the starboard side of the fuselage. This also generated a forward propulsive thrust. In 1948 it set the world speed record at 200 km/h. In 1954–1955 Fairey developed the Jet Gyrodyne, which hovered for the first time in 1957 and into transition in 1968. The final development by Fairey was the Rotodyne – see Figure 1.12. This was a true convertiplane and demonstrated how one airframe could use powered rotors for vertical take-off and landing but perform transitions to and from autogyro-type forward flight.

Figure 1.12   The Rotodyne (Courtesy Agusta Westland)

The 1950s and 1960s decades saw many firsts. In June 1950 the world's first scheduled helicopter passenger service was flown by a British European Airways S51 helicopter. Maiden flights were several: namely, the first turbine helicopter – Kaman K225 in December 1951; the first production rotorcraft with retractable undercarriage – Sikorsky S56 Mojave; the first twin-turbine helicopter – Kaman HTK-1 Huskie; the first Soviet production helicopter – Mi 6 Hook; and the development of the flexstrap rotor hub by Hughes.

In 1962, a Sikorsky S61 was the first helicopter to achieve over 200 mph in level flight; in 1963, the Alouette became the first commercial turbine helicopter in the USA; and in September 1965, the Bell AH1 Cobra became the first dedicated US attack helicopter.

In this period of time several aircraft appeared with the designation XV. In 1955 McDonnell produced the XV1 which was a compound helicopter with the main rotor being based on the Doblhoff concept of the tip jet. It had a wing, a pusher propeller and a twin boom. It achieved flight speeds in excess of 200 mph but suffered from tip-jet noise as did the Fairey Rotodyne.

Sikorsky produced the XV2 in 1952 which had a composite retractable rotor comprising a single blade with a counterbalance weight.

The Bell XV3, see Figure 1.13, of 1955 had a side-by-side rotor layout with each rotor having three blades. It was of the tilt rotor convertiplane configuration. It suffered from long shaft whirl mode instability. In 1958, it was redesigned to have two-bladed rotors with a short shaft.

The XV15 was rolled out in 1976 achieving its first flight in 1979. In 1980 it achieved a flight speed of 300 knots. This aircraft was the forerunner of the V22 which flew for the first time in 1989 and is in extensive use today.

The Kamov Ka22 of 1961 was developed in the Soviet Union and possessed a side-by-side layout. It was unusual in having forward-facing propellers at the wing tips.

In November 1965 we find the Piasecki 16H-1 compound helicopter, see Figure 1.14, making its first flight and achieving 225 mph. This company is still active in the compound helicopter arena. The compound configuration has seen a number of efforts over the years. In the 1960s there was the Sikorsky S65 Blackhawk, a gunship based on the S61 rotor system but with a sleek fuselage shape and fitted with wings. Also in this period there was the Lockheed Cheyenne. Not only was it a compound helicopter with wings and an empennage of intermeshing tail rotor and pusher propeller, but also it had a main rotor controlled by a gyro placed above the main rotor head.

Figure 1.13   Bell XV3 (Wikimedia)

Figure 1.14   Piasecki 16H-1 (Courtesy Piasecki Aircraft Corp.)

We also now begin to see the emergence of extended mission lengths. In 1965, Gary Maher piloted a Hughes 300 to achieve the first solo crossing of the USA. It was a 10 day flight and required 33 refuelling stops. In 1967, a Sikorsky S61 was the first helicopter to achieve the notable goal of crossing the Atlantic Ocean. It took another 15 years (30 September 1982) before a Bell 206-1 became the first helicopter to achieve circumnavigation of the Earth. In July of the following year a Bell Jet Ranger was piloted to the first solo global circumnavigation.

Speed has always been the touchstone of the rotary-wing aircraft and there are many attempts at achieving fast forward flight. Among many, in the UK the most notable of late happened on 11 August 1986. On this date, a Westland Lynx helicopter, flying over the Somerset Levels with Trevor Egginton at the controls, obtained the class world speed record of 400 km/h – see Figure 1.15.

Figure 1.15   Westland Lynx world speed record airframe (Courtesy Agusta Westland)

Figure 1.16   Alouette SA 315 Lama (Courtesy Steve Rod)

Figure 1.17   Eurocopter AS350 in the Himalaya (Courtesy Eurocopter)

The highest altitude was achieved by an Alouette SA315 Lama on 21 June 1972 when it flew to 40 820 ft – see Figure 1.16.

While this remains the altitude record it should not be forgotten that on 14 May 2005 a Eurocopter AS350 helicopter, piloted by Didier Delsalle, was the first to land on the summit of Mount Everest (see Figure 1.17), the highest point on Earth.

Just to prove the point, because of equipment difficulties, it did it again the following day!

1.1.6   The Helicopter from an Engineering Viewpoint

It is easy to invent a flying machine;
more difficult to build one;
to make it fly is everything.
        Otto Lilienthal, 1848–1896

Lilienthal was a pioneer of gliding flight; however, the above quote makes the reader ponder whether he had the helicopter in mind when he wrote it. So far in this chapter, the road to the modern helicopter is anything but straight, so if we examine the quotation line by line, the concept of a lifting rotor constitutes the essential invention. Making it of large radius is simply taking advantage of Newton's second and third laws, which guarantee that in generating a thrust force, by imposing a momentum change on the air, the use of a large quantity of air allows a low-speed change in the air which can be proved to be an efficient way of producing a thrust. When it comes to building the flying machine, the problems of directing it around the sky have to be thought out and translated into hardware: ultimately, however, the solutions for the helicopter are both straightforward and impressive. Upward lift is obtained with the rotor shaft essentially vertical; forward (or backward or sideways) propulsion is achieved by tilting the rotor plane in the desired direction (the rotor shaft itself is tilted in several early designs). This tilting of the rotor disc plane permits moments about the helicopter centre of gravity to be produced, which provides for manoeuvring. Here is a system more elegant in principle than that of a fixed-wing aircraft, where such integration of functions is not possible. However, the combining of several features of rotor control in one function causes its own difficulties.

One can pursue the helicopter rotor's virtues one stage further by noting that the direction of airflow through the rotor becomes reversed in descent allowing blade lift to be produced without power ('autorotation'), permitting a controlled landing in the event of engine failure. This is the point where the heredities of helicopter and autogyro merge and their place on the family tree is defined. These points were made by J.P. Jones in the 1972 Cierva Memorial Lecture to the Royal Aeronautical Society [1]. To quote him at this juncture:

> Can we wonder that the conventional rotor has been a success? At this stage one might think the real question is why the fixed-wing aircraft has not died out.

If we now return to Lilienthal we see the difficulty. Making the helicopter fly involves wrestling with a long catalogue of problems, some of which have been solved while there are others still to be solved – helicopters will always remain an intriguing challenge. During the gestation of the helicopter, it was necessary to invent the use of a tail rotor to stop the helicopter spinning round on the main rotor axis. To this is added other mechanisms of controlling the helicopter in yaw. It took the genius of Juan de la Cierva to devise a system of articulated blades to prevent the aircraft rolling over continuously – his earlier working life involved structures, where the use of a pinned connection to isolate moments provided an ideal grounding. While it can take off, land and hover efficiently, the helicopter can never fly fast judging by fixed-wing aircraft standards, the restriction, surprisingly enough, being one of blade stalling. Climbing is straightforward aerodynamically but descending involves a deliberate venture into the aerodynamicist's nightmare of vortices, turbulence and separated flow. The behaviour of vortices left by an aerodynamic device is crucial to its performance and the locations of these vortices are critical.

As an example, Figure 1.18 shows the location of a typical wake from a fixed-wing aircraft (AlphaJet) where the wing tip vortices stream behind the aircraft and are shown condensed water vapour acting as a tracer.

Figure 1.19 shows vortices being left by the propeller blade tips of an Alenia C27J aircraft and how they wash over the inboard wing structure. To examine a situation in the rotorcraft world, Figure 1.20 shows the vortex wake off a helicopter rotor (AH-1W). Close examination of the wake structure shows interaction with the tail rotor and the rear fuselage.

In addition to influencing any lifting surfaces, vortices interact with each other. Figure 1.21 shows the wake from the wings of a BAE Systems Hawk 200 aircraft after a tight pull-up

Figure 1.18   Tip vortices generated by a fixed-wing aircraft

manoeuvre. The two tip vortices have closed together and on bursting (an unstable breakdown of the initial vortex structure) are beginning to form the characteristic loops associated with parallel pairs of vortices. Figures 1.22a–e show a sequence of images of a small helicopter with a tip-driven rotor. The tip jet uses fuel which produces a considerable amount of water vapour and, using this as a tracer, the exhausts show the wake structures in several different flight regimes.

Figure 1.22a illustrates a hovering condition close to the ground surface. The wake can be seen to contract immediately below the rotor but then expand as the downflow from the rotor is interrupted by the ground forcing it to spill outwards. This phenomenon is called 'ground effect' and is a very important feature of helicopter performance. The wake structure shows not only the 'tube' of vorticity, but also the individual blade tip vortices. Figure 1.22b shows the rotor at low forward speed. Ground effect is still present but the wake is now dispersed rearwards. As the forward speed increases, Figure 1.22c, the vortex 'tube' adopts a sheared profile for a short distance before mutual interaction between the vortices begins to distort the wake. The sheared

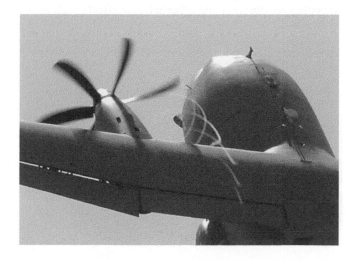

Figure 1.19   Propeller tip vortices

Figure 1.20   Rotor blade tip vortex trajectories (AH-1W) (Courtesy Safety Centre US Navy)

vortex tube concept is a useful modelling technique but, with vortex interactions, as shown in Figure 1.22c, needs care in application. As forward speed increases further, the individual wake vortices show a cycloidal shape (in plan) and a roll-up character at the lateral rotor disc edges, not unlike a fixed wing. These characteristics are well shown in Figures 1.22d and e. In order to provide rotor control, previous discussion has highlighted the necessity of allowing the blades to move out of the plane of rotation. 'Blade articulation' is the term used for the use of conventional rotational hinges and this can lead to comparatively sluggish control since it is really the rotor thrust line change which can impart a turning moment to the helicopter, the free blade attachment hinges contributing very little. This can be significantly improved by adopting the principle of a hingeless rotor where the blade attachment is a flexible component. This can now provide a significant addition to the rotor thrust effect and obtain a greater turning effect for the helicopter. However, this situation is a double-edged sword since the improvement in the path for controlling the helicopter also applies to vibration and other adverse effects such as worsening aircraft

Figure 1.21   Wing tip vortices (BAE Hawk 200)

Figure 1.22   Rotor wake development (Courtesy ATI Corporation): (a) hover in ground effect (IGE); (b) hover IGE to out of ground effect (OGE); (c) low forward flight speed; (d), (e) high forward flight speed

stability. With any practical combination of stability and control characteristics the helicopter remains a difficult and taxing aircraft to fly and generally requires autostabilization to restrict the pilot workload to a safe and comfortable level.

It would seem that we have on our hands a veritable box of tricks. What is certain, however, is that the modern world cannot do without the helicopter. It has become an invaluable asset in many fields of human activity and the variety of its uses continues to increase.

Moreover, to come close to the purpose of this book, the problems that have been solved, or, if only partly solved, at least understood, make good science, high in interest value. This, the book purports to show.

Up to the present, the single-rotor helicopter remains by far the most numerous worldwide and in this book we concentrate exclusively on that type. Its familiar profile, sketched in Figure 1.23, is the result of practical considerations not readily varied. The engines and gearbox require to be grouped tightly around the rotor shaft and close below the rotor. Below them the

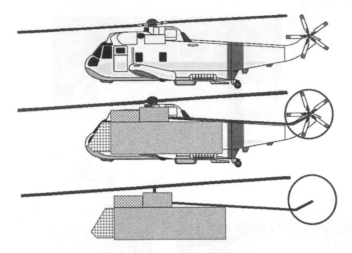

Figure 1.23   The basic structure of a single-rotor helicopter

payload compartment is centrally placed fore and aft to minimize centre of gravity (CG) movements away from the shaft line. In front of the payload compartment is the flight cabin. The transmission line from gearbox to tail rotor needs to be as straight and uninterrupted as possible. Put a fairing around these units so defined and the characteristic profile emerges.

Of the other helicopter configurations, the tandem is the next to consider – a typical example, a Boeing Vertol H46 Sea Knight, is shown in Figure 1.24.

As shown in the figure, there are two rotors, placed at each end of the fuselage. They rotate in opposite senses so the aircraft will respond in yaw to the difference in torques from both of the rotors and a tail rotor is not used. The rear rotor is placed on a pylon so that in normal flight it is not immersed in the wake from the front rotor. This difference in rotation planes gives rise to interactions between translation and rotations during particular flight phases – but these are well known and appropriate action takes place. The figure shows the aircraft decelerating to a hover and to do this the entire aircraft is in a nose-up attitude. As can be seen, the rear rotor is

Figure 1.24   Boeing Vertol H46 Sea Knight (Courtesy US Navy)

Figure 1.25   Mil 12 (Courtesy Agusta Westland)

now moved towards the downwash of the front rotor and this can give rise to a sinking of the rear of the fuselage unless appropriate thrust changes are made. The location of the rotors at each end of the fuselage gives this configuration a wide CG range which for the single main and tail rotor configuration is very limited. The transmission is more complicated as the rotors must be kept in synchronization as the rotor discs overlap. Typically it consists of the engines supplying a combiner gearbox from which individual shafts supply the necessary torque to the rotors.

In the side-by-side configuration, there are a pair of contra-rotating main rotors, but in this configuration they are located laterally on pylons. Yaw control is by differential torque, but for this configuration the CG range is now in a lateral sense. A good example of this type is the Mil 12 shown in Figure 1.25.

This aircraft was designed, in the days of the Cold War, to carry large missiles and the lateral arrangement of the rotors gives a reduced interference between the rotors for an improved performance. The rotors require cross-shafting to keep them synchronized. The Mil 12 was described as appearing to swim through the air with the rotors advancing along the centre looking as if they are executing the breaststroke.

In the coaxial configuration, the rotors are contra-rotating but rotating about a common shaft. An example is the Kamov shown in Figure 1.26.

Figure 1.26   Kamov Ka 27 Helix (Courtesy US Navy)

Figure 1.27   Kaman KMax (Courtesy Marcus Herzig)

As can be seen, the rotor controls have to pass along the same axis so the upper rotor controls have to pass though those of the lower rotor. The rotors are articulated and so the effect of forward speed will tend to tilt the discs rearwards and laterally in opposition so that on one side of the rotor the blade tips are moving together. To avoid any chance of a blade clash, the rotor hubs are separated by a relatively long rotor mast. This gives rise to an increase in drag. With no tail rotor and a compact footprint, they are very useful in shipborne operations.

The helicopter company Kaman has for years used a dual-rotor system which occupies not much more than a rotor disc but uses two rotor shafts inclined to the aircraft's central plane. The most recent of these aircraft is the KMax as shown in Figure 1.27.

The rotors have two blades and are phased by 90° which permits the rotors to pass without mechanical interference. Like the coaxial layout, this compact arrangement permits use in confined spaces. It has therefore contributed much effort in shipborne roles and those with very limited operational areas such as forestry.

The above discussion is concerned with proper helicopter types. If the restriction is lifted then other configurations can be examined, two of which are the tilt rotor and tilt wing.

The classic tilt wing is the Ling–Temco–Vought XC142 as shown in Figure 1.28.

The rotor(s) have been replaced with propellers, two on each wing. The entire wing/engine/propeller layout rotates through 90° so that vertical take-off and landing can be achieved with a conventional lifting through the propeller thrusts and revert to conventional wingborne flight by rotating the wing assembly back to horizontal. With the addition of wing devices such as flaps, the mechanical side of this arrangement is complex. The wing chord is always aligned with the propeller thrust line and so the slipstream can pass smoothly past the upper and lower surfaces. Transition is a difficult flight regime, particularly coming in to land when the wing is positioned with a high pitch angle and the supporting lift is being transferred to the propellers. Wing stall is a potential problem.

The more common tilt rotor variant only tilts the rotor/engine nacelle assemblies leaving the wing in a fixed position relative to the fuselage. The transition from the wing to the rotors is not applicable here, but in and around the hover, the rotor downwash is interrupted by the wing which will be close to a right angle to the flow. This will generate considerable download which

Figure 1.28   LTV XC 142 tilt wing (Courtesy McDermott Library, University of Texas)

is exacerbated by the central parts of the downwash flowing inward along the wings and then forming a fountain flow which creates a larger download still. The use of wide chord flaps aligned vertically reduces the area facing the rotor downwash. A recent example of this type of rotorcraft is the Bell Agusta 609, shown in Figure 1.29.

The final configuration to be highlighted within the helicopter family is the compound helicopter. There have been several companies who have pursued this variant, one of which, Piasecki, has already been mentioned in this chapter. Another landmark move into this type is the AH64A Cheyenne as shown in Figure 1.30.

This helicopter had a gyro-controlled main rotor; the tail rotor was located at the rear of a conventional tail boom alongside a pusher propeller. Stub wings were fitted to the fuselage, which gave this aircraft a fully compounded layout. They also provided an aerodynamic mounting for stores and ordnance.

Any further discussion of vertical take-off and landing (VTOL) configurations now leads into fan and jet lift which is moving away from the idea of rotorcraft and forms a natural halt to this discussion.

Figure 1.29   Bell Agusta BA609 tilt rotor

Figure 1.30   AH64A Cheyenne compound helicopter (Courtesy US Army)

1.2   Book Presentation

It will be helpful to explain certain logistics of the presentation. Symbols are defined when first introduced but for ease of reference are also collected in a list at the start of the book. As concerns units, where there is complete freedom of choice the metric system is preferred; since, however, much use continues to be made of imperial units, particularly in the USA, I have also employed these units freely in numerical examples, sometimes giving both. Again there are tables at the start defining primary and derived units and listing the conversion factors. Lastly, on the question of references, these are numbered in each chapter and listed at the end of the chapter in the usual way. Exception is made, however, in the case of six standard textbooks, which are referred to repeatedly, usually for further information on a topic where the present short treatment is deemed to have gone far enough. The books are:

1. Bramwell, A.R.S. (1976) *Helicopter Dynamics*, Edward Arnold.
2. Johnson, W. (1980) *Helicopter Theory*, Princeton University Press.
3. Stepniewski, W.Z. and Keys, C.N. (1984) *Rotary-wing Aerodynamics*, Vols I and II, Dover.
4. Leishman, J.G. (2006) *Principles of Helicopter Aerodynamics*, 2nd edn, Cambridge Aerospace Series, Cambridge University Press.
5. Padfield, G.D. (2007) *Helicopter Flight Dynamics: The Theory and Application of Flying Qualities and Simulation Modelling*, 2nd edn, Blackwell.
6. Cooke. A and Fitzpatrick, E. (2002) *Helicopter Test and Evaluation*, Blackwell.

In the texts, these are called upon by author's name and no further reference is given.

With this brief introduction we are poised to move into the main treatment of our subject. Finally I would like to mention the following text, which is a source of valuable historical information on helicopters – Boyne, Walter J., Lopez, Donald S. (1984) *Vertical Flight – The Age of the Helicopter*, Smithsonian Institution Press.

Reference

1. Jones, J.P. (1973) The rotor and its future. *Aero. J.*, **751**, 77.

2

Rotor in Vertical Flight: Momentum Theory and Wake Analysis

We begin examining rotor aerodynamics by building models based on momentum transfer – see Glauert [1]. This allows the essential performance of the rotor to be assessed. The most straightforward flight condition is hover, which provides the first part.

2.1 Momentum Theory for Hover

The simplest method that describes the lifting rotor is actuator disc theory. It is based on achieving a lifting force by generating a change of momentum. It assumes the existence of a streamtube which is an axially symmetric surface passing through the rotor disc perimeter which isolates the flow though the rotor. The air is assumed to be incompressible and therefore the flow past any cross-section of the streamtube is constant (Figure 2.1). This also means that because the flow is one dimensional, the flow must remain in the same direction, which for most flight conditions is appropriate. However, this does give rise to a failing of the theoretical model under certain flight conditions.

The flow enters the streamtube, is accelerated through the rotor disc and then is exhausted from the bottom of the streamtube. Far upstream of the disc, the vertical flow velocity must tend to zero making the streamtube cross-section infinite in size. However, the streamtube establishes itself and passes through the rotor disc perimeter.

The additional velocity of V_i as it passes through the rotor is known as the *induced velocity*. It finally forms the wake with a velocity increase of V_2.

The rotor thrust force, T, can be evaluated by considering the momentum increase. The continuity of the flow through the streamtube allows the following:

$$\rho A \cdot V_i = \rho A_2 \cdot V_2 \tag{2.1}$$

Basic Helicopter Aerodynamics, Third Edition. John Seddon and Simon Newman.
© 2011 John Wiley & Sons, Ltd. Published 2011 by John Wiley & Sons, Ltd.

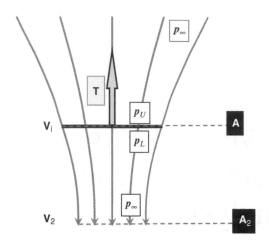

Figure 2.1   Inflow through rotor in hover

The rate of change of momentum gives the rotor thrust as:

$$T = \rho A V_i \cdot V_2 \qquad (2.2)$$

Every second a packet of fluid enters the streamtube with zero vertical velocity. In that same second, an equal packet of fluid leaves the streamtube with a vertical velocity of V_2. Hence, in every second there is a momentum generation given by (2.2).

Actuator disc theory also uses the fact that the thrust can also be expressed in terms of the difference of air pressure on both sides of the rotor disc. In order to generate a thrust, there has to be a pressure difference, which is discontinuous since the rotor disc has zero thickness. However, the airflow through the rotor is continuous and forms part of the theoretical premise of the actuator disc.

We therefore have the following:

$$T = A(p_L - p_U) \qquad (2.3)$$

The final consideration is the application of Bernoulli's equation. This equation can be applied to the flow above or below the rotor disc, but *not* through it. Above the rotor we have:

$$p_\infty = p_U + \frac{1}{2}\rho V_i^2 \qquad (2.4)$$

while below the rotor:

$$p_L + \frac{1}{2}\rho V_i^2 = p_\infty + \frac{1}{2}\rho V_2^2 \qquad (2.5)$$

Subtracting these gives:

$$p_L - p_U = \frac{1}{2}\rho V_2^2 \qquad (2.6)$$

Assembling (2.2), (2.3) and (2.6) gives:

$$V_2 = 2V_i \tag{2.7}$$

In other words, the induced velocity is doubled as the air forms the wake far downstream of the rotor.

Combining (2.2) and (2.7) gives the following result:

$$V_i = \sqrt{\frac{T}{2\rho A}}$$
$$= \sqrt{\frac{1}{2\rho}} \cdot \sqrt{\frac{T}{A}} \tag{2.8}$$

The second expression shows that the induced velocity is dependant explicitly on the disc loading T/A.

Equation 2.8 provides the link between disc loading and induced velocity. At first sight, all this would seem to indicate is that the higher the disc loading, the greater the downdraught from the rotor. Correct, but not the real punchline. The thrust force is working on a medium passing through it at the induced velocity, V_i. Therefore the rotor is expending power (product of force and velocity) given by:

$$P_i = T \cdot V_i \tag{2.9}$$

This power is given a suffix of 'i' consistent with the induced velocity. This is because this power is termed the induced power and is a result of generating the thrust force. If we now compare various rotors, it is apparent that the higher the disc loading, the higher the induced power. As will be shown later, the induced power forms the majority of the power consumed in hover, which is itself a high power-consuming flight regime. The disc loading is therefore one of the first items to be considered when designing a rotor system. Apart from the maximum all-up weight of the helicopter, the main rotor size is almost the first decision to be made.

2.2 Non-dimensionalization

In assessing rotor performance and the ability to compare calculations or different rotors, non-dimensional quantities are useful. The induced velocity is normalized using the rotor tip speed, V_T. The velocity varies along the entire rotor blade but the tip speed is the defining value. In fixed-wing terms this problem does not arise, since the entire wing sees the same velocity. This defines the non-dimensional induced velocity thus:

$$\lambda_i = \frac{V_i}{V_T} \tag{2.10}$$

The thrust force is also normalized in a manner very close to that used for fixed-wing lift – that is, the product of a pressure and an area. The pressure is the dynamic pressure seen at the rotor blade tips (in hover) and the area is the total disc area. A fixed wing uses the planform area which in a helicopter rotor would be the planform area of the blades. While this will be used later, with momentum theory the blades are not considered, so the overall rotor disc area is appropriate. The thrust coefficient is then defined by:

$$C_T = \frac{T}{\frac{1}{2}\rho V_T^2 \cdot A}$$

(2.11)

$$A = \pi R^2$$

The inclusion of the half in the denominator is consistent with the lift coefficient definition for a fixed-wing aircraft. However, the inclusion of the half is not universal. The reader is urged to always check the definition of the thrust coefficient if they are consulting any technical documentation. An error here is both annoying and potentially very serious. Combining (2.8), (2.10), (2.11) gives the following non-dimensional equation:

$$\lambda_i = \frac{1}{2}\sqrt{C_T}$$

(2.12)

$$C_T = 4\lambda_i^2$$

The induced power coefficient is also normalized – noting the inclusion of an extra velocity (tip speed) in the denominator to balance the units:

$$C_{P_i} = \frac{P_i}{\frac{1}{2}\rho V_T^3 \cdot A}$$

(2.13)

Combining (2.9)–(2.13) gives:

$$C_{P_i} = C_T \cdot \lambda_i$$
$$= \frac{1}{2}(C_T)^{3/2}$$

(2.14)

2.3   Figure of Merit

The induced power P_i is the major part of the total power absorbed by a rotor in hover. A further power component is needed, however, to overcome the aerodynamic drag of the blades: this is the profile power P_o, say. Since it is the induced power which relates to the useful function of the rotor – that of producing lift – the ratio of induced power to total power provides a measure of rotor efficiency in the hover. This ratio is called the figure of merit, commonly denoted by M. Using the results of simple momentum theory, M may be variously expressed as:

$$M = \frac{\text{ideal induced power}}{\text{actual induced power}}$$

$$= \frac{P_i}{k_i P_i + P_P} \tag{2.15}$$

Equation 2.15 contains a number of considerations. Firstly, there is the fact that, so far, the induced power is considered ideal. The fixed-wing world has a similar situation where the overall drag of the wing is being considered. The induced drag component has a minimum value when the wing is elliptically loaded. This condition also gives a constant downwash behind the wing. Under the majority of circumstances, a wing is not precisely elliptically loaded and so the downwash will vary across the wing span and an induced drag in excess of the minimum will be encountered. This is catered for by means of a factor applied to the induced drag. With the helicopter rotor, the downwash, or induced velocity, has been assumed constant and so the actuator disc provides the ideal solution. In reality there will be a variation in the induced velocity and so a factor (induced power factor k_i) is applied. The fixed wing also generates a profile drag force due to skin friction, independent of any wing lift. The helicopter rotor also incurs a power requirement to overcome skin friction forces on the blades. The variation of the velocity over the blades requires a short integration to be undertaken, but the profile power is given by:

$$P_P = \frac{1}{8}\rho V_T^3 \cdot NcR \cdot C_{D0} \tag{2.16}$$

The expression contains an area term, which is the blade area. This is because of the way the profile drag is determined by the forces on the blades. In order to align with the induced power normalization the following expression for the profile power coefficient results:

$$C_{PP} = \frac{\frac{1}{8}\rho V_T^3 \cdot NcR \cdot C_{D0}}{\frac{1}{2}\rho V_T^3 \cdot A} \tag{2.17}$$

$$= \frac{1}{4} \cdot \frac{NcR}{A} \cdot C_{D0}$$

Equation 2.17 contains the ratio of blade area to disc area. This is known as the solidity of the rotor and is denoted by s or sometimes σ. The former will be used in this book and can take several forms thus:

$$s = \frac{NcR}{A}$$

$$= \frac{NcR}{\pi R^2} \tag{2.18}$$

$$= \frac{Nc}{\pi R}$$

Equation 2.15 can be rewritten using these non-dimensional quantities thus:

$$M = \frac{C_T \cdot \lambda_i}{k_i C_T \cdot \lambda_i + \frac{s}{4} \cdot C_{D0}}$$

$$= \frac{\frac{1}{2}(C_T)^{3/2}}{k_i \frac{1}{2}(C_T)^{3/2} + \frac{s}{4} \cdot C_{D0}} \tag{2.19}$$

$$= \frac{(C_T)^{3/2}}{k_i \cdot (C_T)^{3/2} + \frac{s}{2} \cdot C_{D0}}$$

Now for a given rotor blade the drag coefficient, and hence the profile power, may be expected not to vary greatly with the level of thrust, provided the blades do not stall nor experience high-compressibility drag rise. Equation 2.19 shows therefore that the value of M for a given rotor will generally increase as C_T increases (*this is illustrated in Figure 2.2*). This feature means that care is needed in using the figure of merit for comparative purposes. A designer may have scope for producing a high value of M by selecting a low blade area such that the blades operate at high lift coefficient approaching the stall, but they need to be sure that the blade area is sufficient for conditions away from hover, such as in high-speed manoeuvre. Again, a comparison of different blade designs – section shape, planform, twist, and so on – for a given application must be made at constant thrust coefficient.

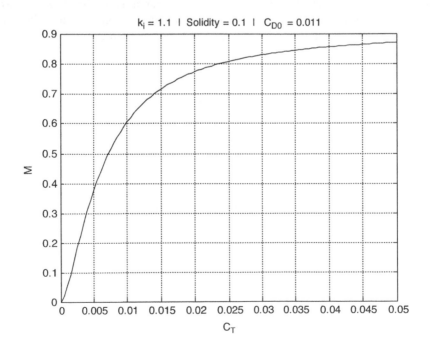

Figure 2.2 Variation of figure of merit with C_T

A good figure of merit is around 0.75, the profile drag accounting for about one-quarter of total rotor power. We may note that for the helicopter as a whole, some power is also required to drive the tail rotor, to overcome transmission losses and to drive auxiliary components: as a result the induced power in hover amounts to 60–65% of the total power absorbed.

2.4 Axial Flight

When the rotor leaves the hovering condition and moves in a vertical sense the flow still remains symmetrical about the thrust force line, that is normal to the rotor disc. In climb, the situation is relatively straightforward to model; however, in descent problems arise. This comment is particularly focused on the use of a momentum-based theory. The flow details become very complex in a medium descent rate condition where the descent rate is of similar magnitude to the induced velocity in hover. A quick view of this type of condition sees opposing flows at the rotor disc of similar magnitude. In order for a momentum theory to be applicable, there must be a realistic throughput of flow along the entire streamtube length. The first situation to be investigated is climb.

2.5 Momentum Theory for Vertical Climb

Consider the rotor in climb, where, again, we observe the flow from the point of view of the rotor. The flow enters the streamtube far upstream of the rotor (because of the climb velocity the streamtube has a finite cross-section), and then passes through the rotor itself, finally passing away from the rotor forming the wake. As a momentum change must be generated, and also making the assertion that the rotor is producing a thrust force in a vertically upward direction, the air will accelerate towards the rotor disc as it approaches from the upstream direction and then accelerate further as it moves downstream into the wake. In order to analyse this situation, the flow velocities are shown in Figure 2.3, the streamtube cross-sectional areas in Figure 2.4 and pressures in Figure 2.5.

The air enters the streamtube with velocity V_C and then acquires an additional velocity of V_i as it passes through the rotor disc. It finally forms the wake with a velocity increase, from V_C, of V_2.

The rotor thrust force, T, can be evaluated by considering the momentum increase. The continuity of the mass flow through the streamtube can be expressed thus:

$$\rho A_1(V_C) = \rho A(V_C + V_i) = \rho A_2(V_C + V_2) \tag{2.20}$$

The rate of change of momentum gives the rotor thrust as:

$$T = \rho A(V_C + V_i)V_2 \tag{2.21}$$

Equation 2.21 represents the mass flow with the velocity increase down the entire length of the streamtube.

As discussed in the hover analysis, the thrust can also be expressed in terms of the difference of air pressure on both sides of the rotor disc – Equation 2.3. Again, the final consideration is the

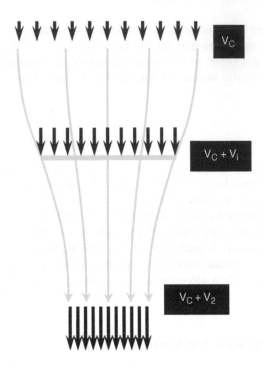

Figure 2.3   Velocity variation

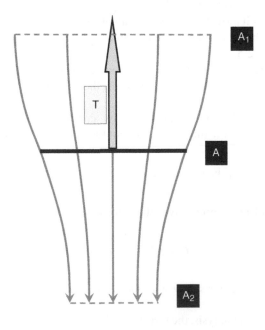

Figure 2.4   Thrust force

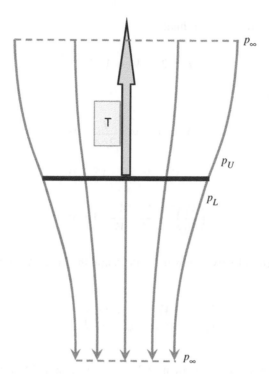

Figure 2.5 Pressure variation

application of Bernoulli's equation. Above the rotor we have:

$$p_\infty + \frac{1}{2}\rho V_C^2 = p_U + \frac{1}{2}\rho(V_C + V_i)^2 \tag{2.22}$$

while below the rotor:

$$p_L + \frac{1}{2}\rho(V_C + V_i)^2 = p_\infty + \frac{1}{2}\rho(V_C + V_2)^2 \tag{2.23}$$

Subtracting these gives:

$$p_L - p_U = \frac{1}{2}\rho(V_C + V_2)^2 - \frac{1}{2}\rho(V_C)^2$$

$$= \frac{1}{2}\rho(2V_C + V_2)V_2 \tag{2.24}$$

Assembling (2.3), (2.21) and (2.24) gives:

$$V_2 = 2V_i \tag{2.25}$$

that is the velocity gain relationship in hover is replicated in climb (*this refers to the velocity increment and not the total velocity*).

Substituting (2.25) into (2.21) we find:

$$T = 2\rho A(V_C + V_i)V_i \tag{2.26}$$

from which we obtain:

$$V_i^2 + V_C \cdot V_i - \frac{T}{2\rho A} = 0$$

$$V_i^2 + V_C \cdot V_i - V_0^2 = 0 \tag{2.27}$$

$$\left(\frac{V_i}{V_0}\right)^2 + \frac{V_C}{V_0} \cdot \frac{V_i}{V_0} - 1 = 0$$

(we denote the induced velocity in hover for the same thrust as V_0) with solution:

$$\frac{V_i}{V_0} = -\frac{V_C}{2V_0} + \sqrt{\left(\frac{V_C}{2V_0}\right)^2 + 1} \tag{2.28}$$

If the climb rate is relatively low, then (2.28) can be expanded to give:

$$\frac{V_i}{V_0} = 1 - \frac{V_C}{2V_0} + \frac{1}{2}\left(\frac{V_C}{2V_0}\right)^2 + \cdots$$

$$= 1 - \frac{1}{2}\left(\frac{V_C}{V_0}\right) + \frac{1}{8}\left(\frac{V_C}{V_0}\right)^2 + \cdots \tag{2.29}$$

The variation in the solution and the various approximations is shown in Figure 2.6.

The first-order solution can be used for small climb rate ratios up to 0.2, the second order can be used to a value of 1.0. Exceeding these limits will give a spurious result.

The power consumed is now given by the product of the thrust and the *total* velocity through the rotor disc, that is:

$$P = T(V_C + V_i)$$

$$= T \cdot V_C + T \cdot V_i$$

$$= P_{\text{CLIMB}} + P_i \tag{2.30}$$

$$\frac{P}{T} = V_C + V_i$$

We have the induced power as before but the climb power is now added.

Considering the last equation of (2.30), the variation of the power thrust ratio (relative to the hover value) is shown in Figure 2.7.

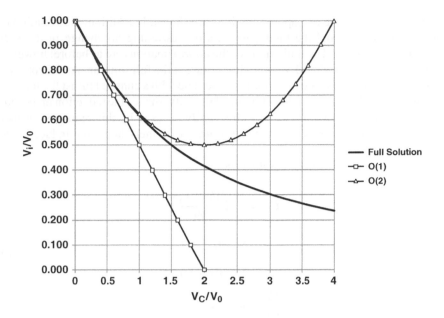

Figure 2.6   Solutions

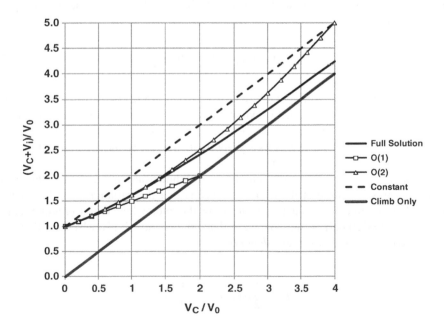

Figure 2.7   Power variation (via velocities)

The full solution and the two approximations are shown. In addition, the variation of the climb-only power ratio and the total power ratio – assuming no change in the induced velocity – are also shown. The limitations of the approximations are as before; however, the climb-only and constant induced-velocity lines bracket the full solution. As the rotor begins to climb, the mass flow entering the rotor increases and with a constant thrust the momentum generation can be achieved with a lower induced velocity – hence the reduction in the induced power. This means that the power to climb steadily is eased by this reduction in induced velocity. In the limit, when the climb rate becomes extremely large, the induced velocity asymptotes to zero.

In non-dimensional terms these results become:

$$
\begin{aligned}
C_P &= C_T\left(\frac{V_C + V_i}{V_T}\right) \\
&= C_T(\mu_z + \lambda_i) \\
&= C_T \cdot \mu_z + C_T \cdot \lambda_i
\end{aligned}
\tag{2.31}
$$

$$
\frac{C_P}{C_T} = \mu_z + \lambda_i
$$

2.6   Modelling the Streamtube

In order to investigate the concept, the velocity variation down the length of the streamtube needs to be modelled. Actuator disc (momentum) theory cannot give a precise solution to this velocity variation; however, a realistic and simple velocity variation can be defined thus:

$$
V = V_C + V_i + V_i \cdot \tanh\left(k\frac{s}{h}\right)
\tag{2.32}
$$

where s is the vertical location variable with the origin at the rotor disc centre – positive downwards; h is a distance defining the extent of the contracting streamtube above and below the rotor disc; and k is a factor which adjusts the severity of the contraction. The hyperbolic tangent function was used because of its asymptotic behaviour so the streamtube finishes as a cylinder far above and below the rotor disc.

Having defined the velocity variation, it is straightforward to determine the pressure variation: the pressure jump at the rotor disc means that the variation must be referenced to either end of the streamtube, far above and below the rotor, where the air pressure returns to ambient. So, using Bernoulli's equation, we find that the pressure is given by the following two expressions (*two are needed as described above*).

Above rotor:

$$
p_\infty + \frac{1}{2}\rho V_C^2 = p + \frac{1}{2}\rho V^2
$$

$$
\frac{p - p_\infty}{\rho} = \frac{V_C^2 - V^2}{2}
\tag{2.33}
$$

Below rotor:

$$p_\infty + \frac{1}{2}\rho(V_C + 2V_i)^2 = p + \frac{1}{2}\rho V^2$$

$$\frac{p - p_\infty}{\rho} = \frac{(V_C + 2V_i)^2 - V^2}{2}$$

(2.34)

If we finally define a pressure coefficient, based on a reference air velocity of U, we find the following results for the pressure variation for above the rotor (C_{PU}) and below the rotor (C_{PL}):

$$C_{PU} = \frac{(V_C)^2 - V^2}{U^2}$$

$$C_{PL} = \frac{(V_C + 2V_i)^2 - V^2}{U^2}$$

(2.35)

with the following values:

V_Z	10 m/s
V_i	10 m/s
R	10 m
U	10 m/s

The velocity, streamtube size and the pressure variation are shown in Figures 2.8–2.10.

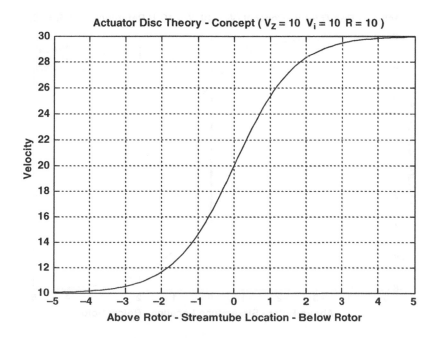

Figure 2.8 Axial velocity variation

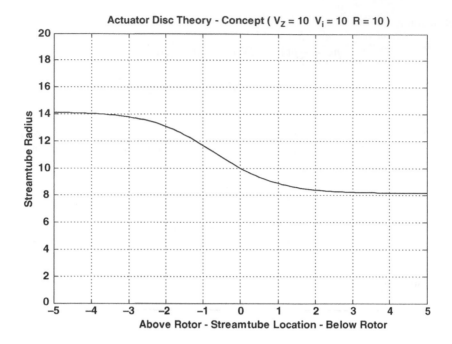

Figure 2.9 Axial streamtube radius variation

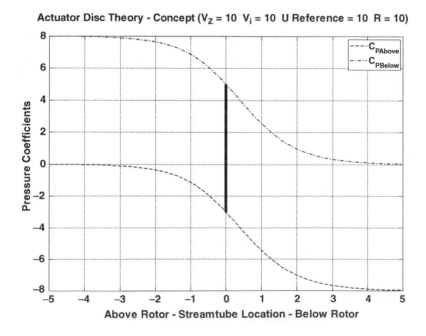

Figure 2.10 Axial pressure variation

The velocity and streamtube radius follow from the foregoing discussion. However, the pressure variation requires some investigation. The pressure line for any location below the rotor lies above that for locations above the rotor. However, it must happen that the overall pressure variation must move, discontinuously, from one solution to the other. It is difficult to justify any movement from one line to another in the free air stream; however, a pressure jump may be successfully argued at the rotor disc. It is this pressure jump that determines the thrust achieved by the rotor and is shown in Figure 2.10.

The above discussion demonstrates the emergence of an actuator disc model which in reality cannot occur. It is not easy to envisage an airflow whose velocity is continuous but also undergoes a discontinuous pressure change at the rotor disc. Therefore one cannot buy an actuator disc; it is purely a conceptual device which describes in fairly good detail the helicopter rotor in axial flight (which includes hover). In the case of climb and hover, this method can be used quite successfully. Unfortunately, this straightforward method has a weakness as it is poor at modelling a descending rotor over a range of descent rates. This will now be investigated.

2.7 Descent

In modelling the actuator disc in descent, the air enters the streamtube from below the rotor with velocity V_D and then acquires a reduction in velocity of V_i as it passes through the rotor disc. It finally forms the wake with a velocity decrease of V_2. In this situation, the upward-going air has its velocity reduced as it passes along the streamtube from below the rotor to above the rotor where the wake forms. This is an effective increase in downward momentum of the air and this is how an upward thrust force can be generated from upward-moving air. The situation is shown in Figures 2.11–2.13.

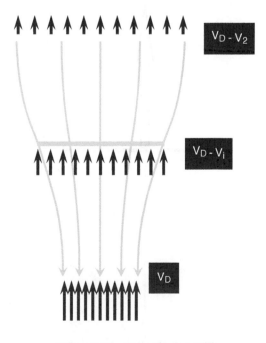

Figure 2.11   Velocity variation

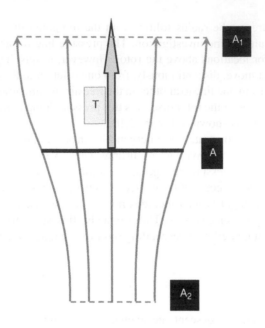

Figure 2.12   Streamtube in descent

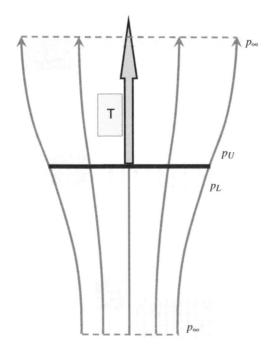

Figure 2.13   Pressure variation

The rotor thrust force, T, can be evaluated by considering this momentum increase. With reference to Figures 2.11 and 2.12, the continuity of the flow through the streamtube can be expressed thus:

$$\rho A_1(V_D - V_2) = \rho A(V_D - V_i) = \rho A_2(V_D) \tag{2.36}$$

The rate of change of momentum gives the rotor thrust as:

$$T = \rho A(V_D - V_i)V_2 \tag{2.37}$$

The thrust can also be expressed in terms of the difference of air pressure on both sides of the rotor disc – see Figure 2.13 – where we have the following:

$$T = A(p_L - p_U) \tag{2.38}$$

As before, Bernoulli's equation can be applied to the flow above or below the rotor disc, but *not* through it. Above the rotor we have:

$$p_\infty + \frac{1}{2}\rho(V_D - V_2)^2 = p_U + \frac{1}{2}\rho(V_D - V_i)^2 \tag{2.39}$$

while below the rotor:

$$p_L + \frac{1}{2}\rho(V_D - V_i)^2 = p_\infty + \frac{1}{2}\rho(V_D)^2 \tag{2.40}$$

Subtracting these gives:

$$\begin{aligned} p_L - p_U &= \frac{1}{2}\rho(V_D)^2 - \frac{1}{2}\rho(V_D - V_2)^2 \\ &= \frac{1}{2}\rho(2V_D - V_2)V_2 \end{aligned} \tag{2.41}$$

Assembling (2.37), (2.38) and (2.41) gives the same result as in climb/hover, namely:

$$V_2 = 2V_i \tag{2.42}$$

from which we obtain from (2.37):

$$T = 2\rho A(V_D - V_i)V_i \tag{2.43}$$

The streamtube modelling which was conducted for climb is now directed at descent. For the descent case, the velocity variation down the length of the streamtube again can be defined relatively simply thus:

$$V = V_D - V_i + V_i \cdot \tanh\left(k\frac{s}{h}\right) \tag{2.44}$$

where the terms are as used in the climb case (*s is still positive downward*).
   The pressure variation now becomes as follows.
   Above rotor:

$$p_\infty + \frac{1}{2}\rho(V_D - 2V_i)^2 = p + \frac{1}{2}\rho V^2$$
$$\frac{p - p_\infty}{\rho} = \frac{(V_D - 2V_i)^2 - V^2}{2} \tag{2.45}$$

Below rotor:

$$p_\infty + \frac{1}{2}\rho V_D^2 = p + \frac{1}{2}\rho V^2$$
$$\frac{p - p_\infty}{\rho} = \frac{V_D^2 - V^2}{2} \tag{2.46}$$

Using the pressure coefficient, based on the reference air velocity of U, we find the following results for the pressure variation for above the rotor (C_{PU}) and below the rotor (C_{PL}):

$$C_{PU} = \frac{(V_D - 2V_i)^2 - V^2}{U^2}$$
$$C_{PL} = \frac{(V_D)^2 - V^2}{U^2} \tag{2.47}$$

   With a descent velocity of 30 m/s, an induced velocity of 10 m/s and a rotor radius of 10 m, the velocity, streamtube size and the pressure variation with axial location are as shown in Figures 2.14–2.16.
   The results are similar to the climb case and the pressure jump required to generate the rotor thrust is again shown.
   An examination of these two analyses might appear to give the impression that descent and climb are closely matched and that there would be no reason to foresee any difficulties. The analyses are beguiling. The sting in the tail is that actuator disc theory assumes a one-dimensional and incompressible flow. Therefore the flow direction must not change throughout the entire length of the streamtube – the constant mass flow guarantees this. In climb it must always be downward. For climb, Equation 2.32 guarantees this will happen. However, for descent, Equation 2.44 admits the possibility of a flow reversal. This begins to define the problems which are faced in modelling the lower descent rate of a helicopter rotor.
   To see the potential problem, recall Equations 2.26 and 2.43. Collating the results and substituting to remove V_2 terms we find the following.
   Climb/hover:

$$T = 2A \cdot \rho(V_C + V_i)V_i \tag{2.48}$$

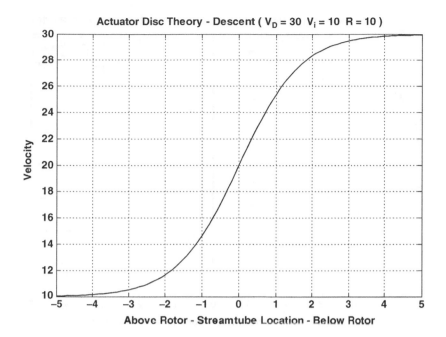

Figure 2.14   Axial velocity variation

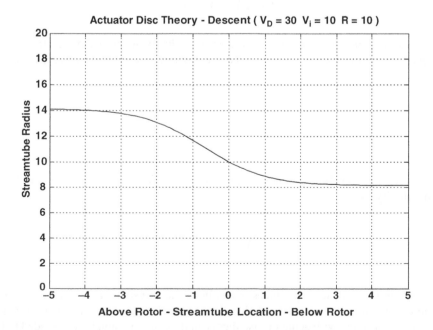

Figure 2.15   Axial streamtube radius variation

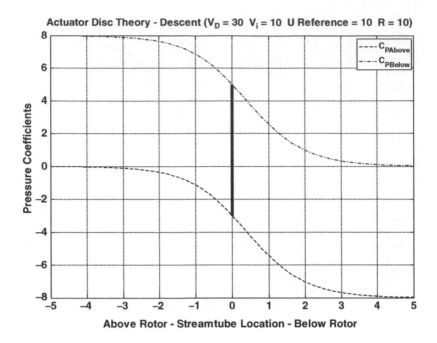

Figure 2.16   Axial pressure variation

Descent:

$$T = 2A \cdot \rho(V_D - V_i)V_i \tag{2.49}$$

If we now set the value of the climb/descent velocity to zero the situation of hover is achieved. Remembering that we denote the hovering induced velocity by V_0, (2.48) and (2.49) become:

$$T = 2\rho A \cdot V_0^2$$
$$T = -2\rho A \cdot V_0^2 \tag{2.50}$$

The upper (climb) equation produces a relationship which sensibly defines the hovering induced velocity as:

$$V_0 = \sqrt{\frac{T}{2\rho A}} \tag{2.51}$$

The second descent equation produces a conflict. The thrust force is always upward and must, therefore, always have a positive value. This cannot happen with this equation. This is an indication of a problem with actuator disc theory in descent – it cannot be extended to hover, leaving a domain which this equation cannot model. As will be seen, the theory works for appropriate values of descent rate but, as just described, it cannot be extended back to the hovering condition.

Combining the results (2.51) with (2.48) and (2.49) gives:

$$V_0^2 = \frac{T}{2\rho A} = (V_C + V_i)V_i$$

$$V_0^2 = \frac{T}{2\rho A} = (V_D - V_i)V_i$$

(2.52)

If we define the following normalized velocity terms:

$$\overline{V_C} = \frac{V_C}{V_0}$$

$$\overline{V_D} = \frac{V_D}{V_0}$$

$$\overline{V_i} = \frac{V_i}{V_0}$$

(2.53)

and make the substitution:

$$\overline{V_D} = -\overline{V_C}$$

(2.54)

(so that both climb and descent use a common velocity sign convention), we obtain the following non-dimensional equations:

$$(\overline{V_C} + \overline{V_i})\overline{V_i} = 1$$

$$(\overline{V_C} + \overline{V_i})\overline{V_i} = -1$$

(2.55)

Equations 2.55 are then solved to give solutions for the induced velocity; however, only positive solutions are physically appropriate. A simple interpretation of these solutions can be obtained by re-expressing Equations 2.55 as:

$$\overline{V_C} = \pm \frac{1}{\overline{V_i}} - \overline{V_i}$$

(2.56)

These represent the sum of a rectified rectangular hyperbola and a linear function as shown in Figure 2.17.

There is the requirement for the air to flow in one direction only (*which was introduced earlier*), and in order to satisfy this condition, limitations on the solutions must be made. Observation of the solutions shows that in order for the flow to be in the same direction at the entry, rotor disc and exit cross-sections of the streamtube, a region of axial velocity must be removed as shown in Figure 2.18.

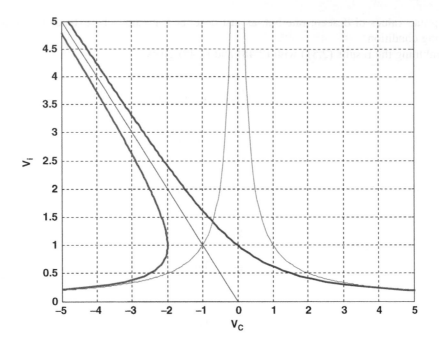

Figure 2.17   Solutions in axial flight

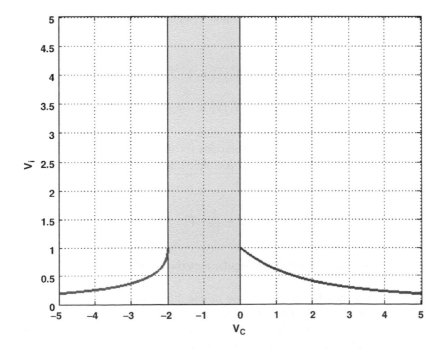

Figure 2.18   Actuator disc theory non-qualifying region

Where the limits on the vertical velocity are those which give a real mathematical solution, Equation 2.55 now becomes:

$$\left(\overline{V_C} + \overline{V_i}\right)\overline{V_i} = 1 \Leftrightarrow \overline{V_C} \geq 0$$
$$\left(\overline{V_C} + \overline{V_i}\right)\overline{V_i} = -1 \Leftrightarrow \overline{V_C} \leq -2$$

(2.57)

It is therefore necessary to seek alternative solutions to the conditions experienced in this region. While this is relatively easy to write down, it is very difficult to handle theoretically. This is because the actuator disc relies on a definable streamtube while the flow conditions in the non-qualifying region are not in any way conducive to such a concept. Indeed, the actual situation is of a system of vortices being generated by the lifting rotor blades and becoming the dominant feature. In axial descent, at low speed, the downwash induced by the rotor and wake is matched by the upward motion induced by the descending rotor, resulting in the vorticity in the wake remaining in close proximity to the rotor disc. It is reasonable to observe that the rotor cannot store vorticity for ever and so there must some manner in which it can disperse. In reality it tends to collect around the rotor giving rise to difficulties in handling and to act as a source of high-frequency vibration. Periodically, this collected vorticity releases, freeing up the rotor, whence the process can start again. This will contribute to the vibration by adding a low-frequency component to the previously mentioned high frequency. As can be envisaged, the airflow characteristics of a helicopter rotor, in what is termed the vortex ring state, are very complex and difficult flow conditions to model theoretically and have exercised many minds over the years.

Although these flight conditions are complex, it is instructive to consider what the mean flow behaviour would be. This directs the discussion to the various flow states that a helicopter rotor can experience as it moves from high-rate vertical climb to high-rate vertical descent.

These are presented schematically in Table 2.1.

2.8   Wind Tunnel Test Results

To examine the real effects of descent rate on the vortex wake streaming from a rotor the results of a wind tunnel test on a model rotor are presented.

The model installation is shown in Figure 2.19.

A particle image velocimetry (PIV) study was performed which enables laser light to be used to determine the flow velocity components in a geometric plane which is that swept out by the laser light beam – see Figure 2.20. In this experiment, the plane is normal to the rotor disc plane and includes the outer end of the blade and the initial part of the vortex wake. The velocity data are processed to give the vorticity variation across a part of the plane. Table 2.2 shows the vorticity maps for a range of wind speeds, which are effectively the descent rate of the rotor. For scaling purposes, the hovering induced velocity takes the value of 1.1 m/s.

The rotor rig details are:

Rotor speed	1200 RPM
Rotor radius	254 mm
Blade chord	34 mm
Number of blades	3
Blade twist	8°
Collective pitch	10°

Table 2.1 Rotor flow states in axial flight

In climb and hover the flow retains the streamtube concept. Throughout the length of the streamtube, the velocity is in a downward direction

Actuator disc theory can be applied with due regard to its inherent simplicity. This is known as the normal working state

At low rates of descent a toroidal type of vortical structure begins to form around the tip region. Actuator disc theory can be applied but this is the sensible limit of its use. There is a small amount of upward flow, relative to the rotor, but the majority is in a downward direction still

At moderate rates of descent – equivalent to the induced velocity in hover – the rotor becomes immersed in a large toroidal vortex type of structure. As already explained, this is a very unsteady type of flow state. Actuator disc theory is totally unsuitable for this situation. This is known as the vortex ring state. The velocity direction is varied across the rotor throughout this flow state

As the descent speed increases, the wake now begins to move above the rotor. Its characteristics are similar to the wake behind a circular disc, hence its name – the turbulent wake state. There is still a difference in velocity direction. It is difficult to justify using actuator disc theory unless the descent velocity is approaching twice that of the induced velocity in hover

As the descent rate increases further, the wake is now moved further above the rotor and the velocity is now upward throughout. The streamtube concept is now appropriate. Actuator disc theory can be applied with due regard to its inherent simplicity. This is known as the windmill brake state

The illustrations show the gradual transition from a hover flow state to that of a descending windmill brake state. The character is a balance of clean wake flow at each end of the table and that of a dispersing wake where the vortex ring state region is found. It should be emphasized that the illustrations are a mean flow. The vortex ring state is a very unsteady flight regime and this must be borne in mind.

Figure 2.19   Rotor rig in wind tunnel

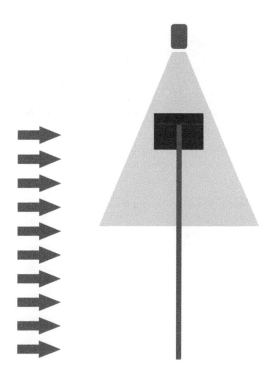

Figure 2.20   Schematic of PIV measurement window

Table 2.2 Vorticity maps for a range of axial wind speeds

V_Z/V_0	PIV image	V_Z/V_0	PIV image
0		3.0	
The rotor wake is clearly defined			
0.5		3.4	
The far wake is showing evidence of dispersion		The wake dispersion is enveloping the flow around the rotor tip region – but descending	
1.0		3.8	
		The wake dispersion is enveloping the flow around the rotor tip region – but ascending	
1.6		4.0	
The wake dispersion is becoming more spread			

Table 2.2 (*Continued*)

V_Z/V_0	PIV image	V_Z/V_0	PIV image
2.0		4.4 The flow is at the point of ascending above the rotor	
2.4		4.8	
2.8		5.0 The flow is ascending and the dispersion is reducing	

2.9   Complete Induced-Velocity Curve

2.9.1   *Basic Envelope*

It is of interest to know how the induced velocity varies through all the phases of axial flight. For the vortex ring and turbulent wake states, where momentum theory fails, information has been obtained from measurements in flight, supported by wind tunnel tests [2–5]. Obviously the making of flight tests (measuring essentially the rate of descent and control angles) is both difficult and hazardous, especially where the vortex ring state is prominent, and not surprisingly the results (see Figure 2.21) show some variation. Nevertheless the main trend has been ascertained and what is effectively a universal induced-velocity curve can be defined.

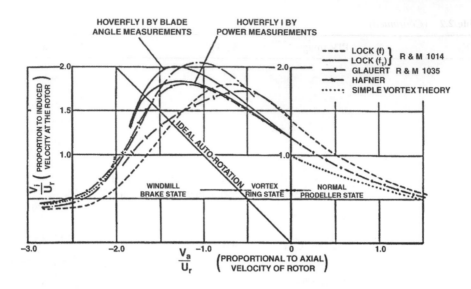

Figure 2.21   Experimental test results for axial flight

This is shown in Figure 2.22, using the simple momentum theory results of Equation 2.55 in the regions to which they apply. We see that on moving from hover into descent the induced velocity increases more rapidly than momentum theory would indicate. The value rises, in the vortex ring state, to about twice the hover value, then falls steeply to about the hover value at entry to the windmill brake state.

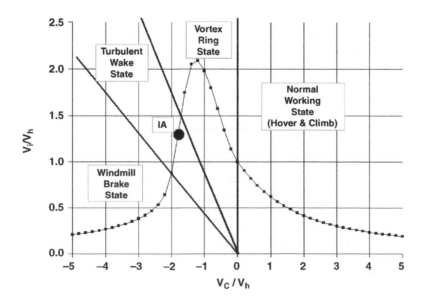

Figure 2.22   Complete induced-velocity curve

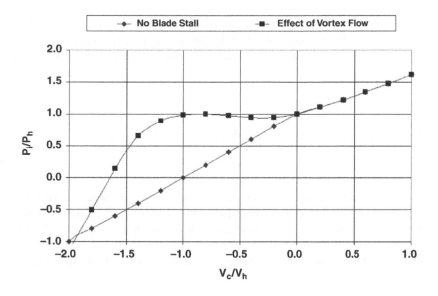

Figure 2.23   Power ratio variation with axial velocity

The power required to maintain thrust in vertical descent generally falls as the rate of descent increases, except that in the vortex ring state an increase is observed (Figure 2.23). The effect appears to be caused by stalling of the blade sections during the violent vortex-shedding action. The increase can be potentially hazardous when making a near-vertical landing approach under conditions in which the engine power available is relatively low, as would be the case under high helicopter load in a high ambient temperature.

2.9.2   Autorotation

Equation 2.30 shows the power required to climb and generate downwash. To this is added the profile power. If we examine the equivalent power consumption in descent we find the following equation:

$$
\begin{aligned}
P &= T(-V_D + V_i) + P_P \\
&= -T \cdot V_D + T \cdot V_i + P_P
\end{aligned}
\tag{2.58}
$$

The profile power expression is independent of the climb/descent speed so the hover result (Equation 2.16) is applicable here. Equation 2.58 now has the possibility of attaining the value zero. This would mean that *no* power input to the engine is required. Assuming this to be the situation, we have the following expression:

$$
V_D = V_i + \frac{P_P}{T}
\tag{2.59}
$$

While this is the simplest form, with no factors applied, it illustrates the importance of disc loading – hence downwash – in the descent rate that is required to establish this condition of no power input. This is the autorotation condition used by helicopters to land safely when power to the main rotor is lost. It is not surprising that this autorotation speed is kept as low as possible, so the downwash should be limited in value. This places a minimum limit on the main rotor radius.

2.9.3   Ideal Autorotation

The point of intersection of the induced-velocity curve with the line:

$$V_C + V_i = 0 \tag{2.60}$$

is of particular interest because it defines what is termed the state of ideal autorotation[1] (IA in Figure 2.22), in which, since there is no mean flow through the rotor, the induced power is zero.

In round terms, values of V_C/V_0 for ideal and real autorotation are about -1.7 and -1.8, respectively.

Pursuing the analogy of flow past a solid plate (turbulent wake state), the plate drag may be written:

$$D = \frac{1}{2}\rho V_C^2 \cdot A \cdot C_D \tag{2.61}$$

and if this is equated to rotor thrust we have:

$$T = 2\rho A V_0^2 = \frac{1}{2}\rho V_C^2 \cdot A \cdot C_D \tag{2.62}$$

from which we find:

$$C_D = \frac{4}{\left(\dfrac{V_C}{V_0}\right)^2} \tag{2.63}$$

Examining Figure 2.24, we see that with $V_C/V_0 = -1.7$, C_D has the value 1.38 which is close to that for a solid plate. A slightly better analogy is obtained by taking the real value $V_C/V_0 = -1.8$, which yields a C_D value of 1.23, close to the effective drag coefficient of a parachute.

Thus in autorotative vertical descent the rotor behaves like a parachute.

2.10   Summary Remarks on Momentum Theory

The place of momentum theory is that it gives a broad understanding of the functioning of the rotor and provides basic relationships for the induced velocity created and the power required in

---

[1] Autorotation is an extremely important facility because in case of a power failure the rotor can continue to produce a thrust approximately equal to the aircraft's weight, allowing a controlled descent to ground to be made. The term 'ideal autorotation' is used because in practice power is still needed to overcome the drag of the blades (profile power).

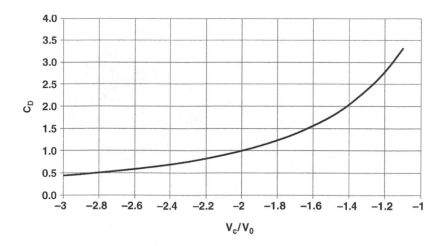

Figure 2.24 Effective flat plate drag

producing a thrust to support the helicopter. The actuator disc concept, upon which the theory is based, is most obviously fitted to flight conditions at right angles to the rotor plane, that is to say the hover and axial flight states we have discussed. Nevertheless, further reference to the theory will be made when discussing forward flight (Chapter 5).

Momentum theory brings out the importance of disc loading as a gross parameter; it cannot, however, look into the detail of how the thrust is produced by the rotating blades and what design criteria are to be applied to them. For such information we need additionally a blade element theory, corresponding to aerofoil theory in fixed-wing aerodynamics. We shall turn to this in Chapter 3.

2.11 Complexity of Real Wake

The actuator disc concept, taken together with blade element theory, serves well for the purposes of helicopter performance calculation. When, however, blade loading distributions or vibration characteristics are required for stressing purposes, it is necessary to take into account the real nature of flow in the rotor wake. This means abandoning the disc concept and recognizing that the rotor consists of a number of discrete lifting blades, carrying (bound) vorticity corresponding to the local lift at all points along the span. Corresponding to this bound vorticity, a vortex system must exist in the wake (Helmholtz's theorem) in which the strength of wake vortices is governed by the rate of change of circulation along the blade span. If for the sake of argument this rate could be made constant, the wake for a single rotor blade in hover would consist of a vortex sheet of constant spanwise strength, descending in a helical pattern at constant velocity, as illustrated in Figure 2.25. The situation is analogous to that of elliptic loading with a fixed wing, for which the induced drag (and hence the induced power) is a minimum. This ideal distribution of lift, however, is not realizable for the rotor blade, because of the steadily increasing velocity from root to tip.

The most noticeable feature of the rotor blade wake in practice is the existence of a strong vortex emanating from the blade tip where, because the velocity is highest, the rate of change of lift is greatest. In hover, the tip vortex descends below the rotor in a helical path.

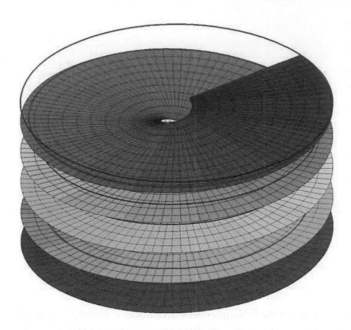

Figure 2.25   Helical wake

This can be visualized in a wind tunnel using smoke injection (Figure 2.26) or other means and is often observable in open flight under conditions of high load and high humidity. An important feature which can be seen in Figure 2.27 is that on leaving the blade the tip vortex initially moves inwards towards the axis of rotation and stays close under the disc plane; in consequence the next tip to come round receives an upwash, increasing its effective incidence and thereby intensifying the tip vortex strength.

Figure 2.28 due to J.P. Jones [6] shows a calculated spanwise loading for a Wessex helicopter blade in hover and indicates the tip vortex position on successive passes. The kink in loading distribution at 80% span results from this tip vortex pattern, particularly from the position of the immediately preceding blade.

The concentration of the tip vortex can be reduced by design changes such as twisting the tip nose-down, reducing the blade tip area or special shaping of the planform, but it must be borne in mind that the blade does its best lifting in the tip region where the velocity is high.

Since blade loading increases from the root to near the tip (Figure 2.28), the wake may be expected to contain some inner vorticity in addition to the tip vortex. This might appear as a form of helical sheet akin to that of the illustration in Figure 2.25, though generally not of uniform strength. Definitive experimental studies by Gray [7], Landgrebe [8] and their associates have shown this to be the case. Thus the total wake comprises essentially the strong tip vortex and an inner vortex sheet, normally of opposite sign. The situation as established by Gray and Landgrebe is pictured, in a diagram which has become standard, by Bramwell (p. 117) and other authors.

Figure 2.29 is a modified version of this diagram, intended to indicate the inner vorticity sheet, emanating from the bound vorticity on the inner part of the blade.

Figure 2.26   Real rotor wake

The Gray and Landgrebe studies show clearly the contraction of the wake immediately below the rotor disc. Other features which have been observed are that the inner sheet moves downwards faster than the tip vortex and that the outer part of the sheet moves faster than the inner part, so the sheet becomes increasingly inclined to the rotor plane.

2.12   Wake Analysis Methods

By analysis of his carefully conducted series of smoke-injection tests, Landgrebe [9] reduced the results to formulae giving the radial and axial coordinates of a tip vortex in terms of azimuth angle, with corresponding formulae for the inner sheet. From these established vortex positions (the so-called prescribed wake) the induced velocities at the rotor plane may be calculated. The method belongs in a general category of prescribed-wake analysis, as do earlier analyses by Prandtl, Goldstein and Theodorsen, descriptions of which are given by

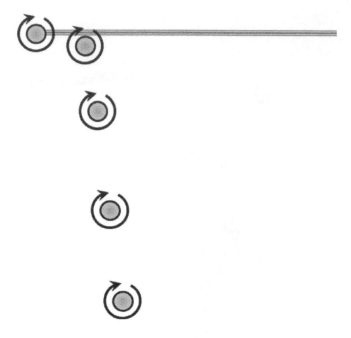

Figure 2.27   Schematic of vortex motion

Bramwell. These earlier forms treated either a uniform vortex sheet as pictured in Figure 2.25 or the tip vortex in isolation, and so for practical application are effectively superseded by Landgrebe's method.

More recently, considerable emphasis has been placed on free-wake analysis, in which modern numerical methods are used to perform iterative calculations between the induced-velocity distribution and the wake geometry, both being allowed to vary until mutual

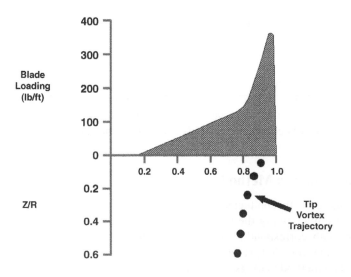

Figure 2.28   Blade loading

Figure 2.29   Wake sheet and tip vortex trajectory

consistency is achieved. This form of analysis has been described for example by Clark and Leiper [10]. Generally the computing requirements are very heavy, so considerable research effort also goes into devising simplified free-wake models which will reduce the computing load. The computing power available is consistently increasing with time; however, the complexity of the methods always seems to fill the available computing capability.

Calculations for a rotor involve adding together calculations for the separate blades. Generally this is satisfactory up to a depth of wake corresponding to at least two rotor revolutions. A factor which helps this situation is the effect on the tip vortex of the upwash ahead of the succeeding blade – analogous to the upwash ahead of a fixed wing. The closer the spacing between blades, the stronger the effect from a succeeding blade on the tip vortex of the blade ahead of it; thus it is observed that when the number of blades is large, the tip vortex remains approximately in the plane of the rotor until the succeeding blade arrives, when it is convected downwards. In the 'far' wake, that is beyond a depth corresponding to two rotor revolutions, it is sufficient to represent the vorticity in simplified fashion; for example, free-wake calculations can be simplified by using a succession of vortex rings, the spacing of which is determined by the number of blades and the mean local induced velocity. Eventually in practice both the tip vortices and the inner sheets from different blades interact and the ultimate wake moves downwards in a confused manner.

There we leave this brief description of the real wake of a hovering rotor and the methods used to represent it. This branch of the subject is often referred to as vortex theory. It will be touched on again in the context of the rotor in forward flight (Chapter 5). For more detailed accounts, the reader is referred to the standard textbooks and the more specific references which have been given in these past two sections.

2.13   Ground Effect

The induced velocity of a rotor in hover is considerably influenced by the near presence of the ground. At the ground surface the downward velocity in the wake is, of course, reduced to zero and this effect is transferred upwards to the disc through pressure changes in the wake, resulting in a lower induced velocity for a given thrust. This is shown in Figures 2.30 and 2.31.

The two images illustrate the wake impinging on the water during air sea rescue operations. The outward motion of the waves shows how the vertical velocity at the rotor disc is turned to a horizontal direction by the effect of the sea. The induced power is therefore lower, which is to say that a helicopter at a given weight is able to hover at lower power thanks to 'support' given by the ground. Alternatively put, for a given power output, a helicopter 'in ground effect' is able to hover at a greater weight than when it is away from the ground. As Bramwell has put it, '*the improvement in performance may be quite remarkable; indeed some of the earlier, under-powered, helicopters could hover only with the help of the ground*'.

The theoretical approach to ground effect is, as would be expected, by way of an image concept. A theory by Knight and Hafner [11] makes two assumptions about the normal wake:

Figure 2.30   Sea King hovering close to the ground surface

Figure 2.31 Hovering close to water surface showing wake impingement (Courtesy US Navy)

1. That circulation along the blade is constant, thus restricting the vortex system to the tip vortices only.
2. That the helical tip vortices form a uniform vortex cylinder reaching to the ground.

The ground plane is then represented by a reflection of this system, of equal dimensions below the plane but of opposite vorticity, ensuring zero normal velocity at the surface. The induced velocity at the rotor produced by the total system of real and image vortex cylinders is calculated and hence the induced power can be derived as a function of rotor height above the ground.

It is found that the power, expressed as a proportion of that required in the absence of the ground, is as low as 0.5 when the rotor height to rotor radius is about 0.4, a typical value for the point of take-off. Since induced power is roughly two-thirds of total power (Section 2.1), this represents a reduction of about one-third in total power. By the time the height to radius ratio reaches 2.0, the power ratio is close to 1.0, which is to say ground effect has virtually disappeared. The results are only slightly dependent on the level of thrust coefficient.

Similar results have been obtained from tests on model rotors, measuring the thrust that can be produced for a given power. A useful expression emerges from a simple analysis made by Cheeseman and Bennett [12], who give the approximate relationship:

$$\frac{T}{T_\infty} = \frac{1}{1 - \left(\frac{1}{4Z/R}\right)^2} \tag{2.64}$$

where T is the rotor thrust produced in ground effect and T_∞ is the rotor thrust produced out of ground effect at the same level of power. Z is the rotor height above the ground and R is the rotor radius. The variation is shown in Figure 2.32.

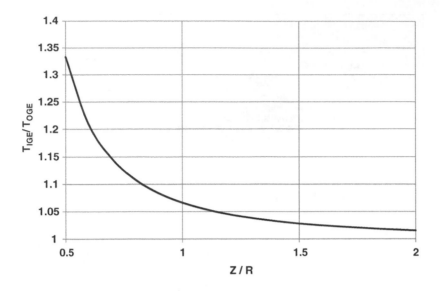

Figure 2.32   Ground effect on rotor thrust

This shows good agreement with experimental data.

Ground effect has a profound influence on a helicopter's performance and so a technical specification will often include two values for a quantity such as power or thrust either in ground effect (IGE) or out of ground effect (OGE).

2.14   Brownout

As the helicopter approaches the ground, the spreading wake will interact with any horizontal wind. This will tend to turn the upwind wake upwards after which it re-enters the rotor. This sets up a type of toroidal flow. The effect is shown schematically in Figure 2.33.

A considerable amount of research has been devoted to this effect, for the reason that when the helicopter is operating where the ground surface is easily raised into the air [13–15]. A cloud of ground debris is entrained by the interacting flows surrounding the aircraft. Visibility is severely restricted making operations very hazardous. Figure 2.34 shows a V22 descending into a cloud of dust caused by the downwash from the rotors.

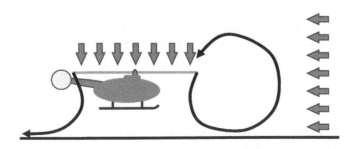

Figure 2.33   Schematic of brownout

Figure 2.34 V22 descending into a brownout cloud (Courtesy Eglin Air Force Base)

References

1. Glauert, H. (1937) *The Elements of Aerofoil and Airscrew Theory*, Cambridge University Press.
2. Gustafson, F.B. and Gessow, A. (1945) Flight tests on the Sikorsky HNS-1 (Army YR-4B) helicopter, NACA MR L5D09a.
3. Gessow, A. (1948) Effect of rotor blade twist and planform taper on helicopter hovering performance, NACA Technical Notice 1542.
4. Brotherhood, P. and Stewart, W. (1949) An experimental investigation of the flow through a helicopter rotor in forward flight, ARC R&M 2734.
5. Castles, W. Jr and Gray, R.B. (1951) Empirical relation between induced velocity, thrust and rate of descent of a helicopter rotor as determined by wind tunnel tests on four model rotors, NACA TN 2474, October.
6. Jones, J.P. (1973) The rotor and its future. *Aero. J.*, **751**, 77.
7. Gray, R.B. (1956) An aerodynamic analysis of a single-bladed rotor in hovering and low speed forward flight as determined from smoke studies of the vorticity distribution in the wake, Princeton University Aeronautical Engineering Report 356.
8. Landgrebe, A.J. (1972) The wake geometry of a hovering helicopter rotor and its influence on rotor performance. *JAHS*, **17**(4), 3–15.
9. Landgrebe, A.J. (1971) Analytical and experimental investigation of helicopter rotor hover performance and wake geometry characteristics, USAAMRDL Technical Report, 71-24.
10. Clark, D.R. and Leiper, A.C. (1970) The free wake analysis – a method for the prediction of helicopter rotor hovering performance. *JAHS*, **15**(1), 3–12.
11. Knight, M. and Hafner, R.A. (1941) Analysis of ground effect on the lifting airscrew, NACA TN 835.
12. Cheeseman, I.C. and Bennett, W.E. (1955) The effect of the ground on a helicopter rotor, ARC R & M 3021.
13. Phillips, C., Kim, H.W. and Brown, R.E. (2009) The effect of rotor design on the fluid dynamics of helicopter brownout. 35th European Rotorcraft Forum, Hamburg, September.
14. Phillips, C., Kim, H.W. and Brown, R.E. (2010) Helicopter brownout – can it be modelled? RAeS Rotorcraft Group Conference 'Operating Helicopters Safely in a Degraded Visual Environment', London, June.
15. Phillips, C., Kim, H.W. and Brown, R.E. (2010) The flow physics of helicopter brownout. 66th American Helicopter Society Annual Forum, Phoenix, AZ, May.

3

Rotor in Vertical Flight: Blade Element Theory

3.1 Basic Method

Blade element theory is basically the application of the standard process of aerofoil theory to the rotating blade. A typical aerodynamic strip is shown in Figure 3.1 and the appropriate notation for a typical strip is shown in Figure 3.2. Although in reality flexible, a rotor blade is assumed throughout to be rigid, the justification for this lying in the fact that at normal rotation speeds the outward centrifugal force is the largest force acting on a blade and, in effect, is sufficient to hold the blade in rigid form. In vertical flight, including hover, the main complication is the need to integrate the elementary forces along the blade span. Offsetting this, useful simplification occurs because the blade incidence and induced flow angles are normally small enough to allow small-angle approximations to be made.

Figure 3.3 is a plan view of the rotor disc, seen from above. Blade rotation is anticlockwise (the normal system in the UK and the USA) with angular velocity Ω. The blade radius is R, the tip speed therefore being ΩR, alternatively written as V_T. An elementary blade section is taken at radius r, of chord length c and spanwise width dr. Forces on the blade section are shown in Figure 3.4. The flow seen by the section has velocity components Ωr in the disc plane and $(V_C + V_i)$ perpendicular to it. The resultant of these is:

$$U = \sqrt{\left[(V_C + V_i)^2 + (\Omega r)^2 \right]}$$

(3.1)

The blade pitch angle, determined by the pilot's collective control setting (see Chapter 4), is θ. The angle between the flow direction and the plane of rotation, known as the inflow angle, is ϕ, given by:

$$\phi = \tan^{-1} \left[\frac{(V_C + V_i)}{\Omega r} \right]$$

(3.2)

Basic Helicopter Aerodynamics, Third Edition. John Seddon and Simon Newman.
© 2011 John Wiley & Sons, Ltd. Published 2011 by John Wiley & Sons, Ltd.

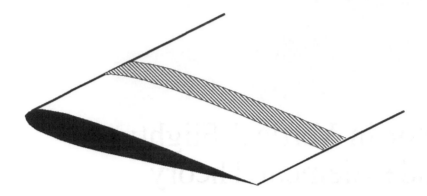

Figure 3.1   General strip used in blade aerodynamic calculations

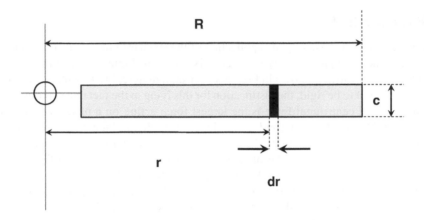

Figure 3.2   Blade strip coordinates

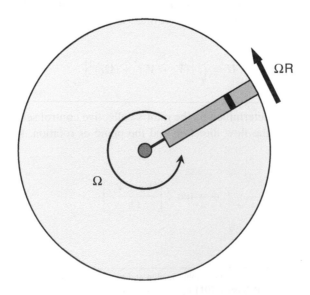

Figure 3.3   Rotor disc viewed from above

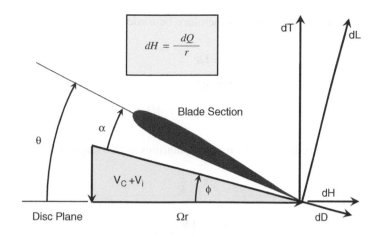

$$dH = \frac{dQ}{r}$$

Figure 3.4   Blade section flow conditions in vertical flight

or for small angles, which we shall assume:

$$\phi = \frac{(V_C + V_i)}{\Omega r} \tag{3.3}$$

The angle of incidence of the blade section, denoted by α, is seen to be:

$$\alpha = \theta - \phi \tag{3.4}$$

The elementary lift and drag forces on the section are:

$$dL = \frac{1}{2}\rho U^2 \cdot c \, dr \cdot C_L$$

$$dD = \frac{1}{2}\rho U^2 \cdot c \, dr \cdot C_D \tag{3.5}$$

Resolving these normal and parallel to the disc plane gives an element of thrust:

$$dT = dL \cos \phi - dD \sin \phi \tag{3.6}$$

and an element of blade torque:

$$dQ = (dL \sin \phi + dD \cos \phi)r \tag{3.7}$$

The inflow angle ϕ may generally be assumed to be small; from Equation 3.3 this may be questionable near the blade root where Ωr is small, but there the blade loads are themselves small also. Masking the reasonable assumption that the rotor blade section has a high lift/drag ratio, the following approximations can therefore be made:

$$U \simeq \Omega r$$

$$dT \simeq dL \tag{3.8}$$

$$dQ \simeq (\phi \, dL + dD)r$$

It is convenient to introduce dimensionless quantities at this stage. We write:

$$x = \frac{r}{R} \tag{3.9}$$

$$\frac{U}{\Omega R} = \frac{\Omega r}{\Omega R} = x \tag{3.10}$$

$$dC_T = \frac{dT}{\frac{1}{2}\rho(\Omega R)^2 \cdot A} \tag{3.11}$$

$$dC_Q = \frac{dQ}{\frac{1}{2}\rho(\Omega R)^2 \cdot A \cdot R} \tag{3.12}$$

$$dC_P = \frac{dP}{\frac{1}{2}\rho(\Omega R)^3 \cdot A} \tag{3.13}$$

$$\lambda = \mu_{ZD} = \mu_Z + \lambda_i = \frac{(V_C + V_i)}{\Omega R} = x\phi \tag{3.14}$$

λ is known as the inflow factor. Now the element of thrust becomes, noting that we have N blades:

$$dC_T = N\frac{\frac{1}{2}\rho U^2 \cdot c \, dr \cdot C_L}{\frac{1}{2}\rho(\Omega R)^2 \cdot A} = N\frac{\frac{1}{2}\rho(\Omega r)^2 \cdot c \, dr \cdot C_L}{\frac{1}{2}\rho(\Omega R)^2 \cdot \pi R^2} = \frac{Nc}{\pi R}C_L \cdot x^2 \, dx \tag{3.15}$$

The definition of thrust coefficient contains the rotor disc area; however, when blade element theory is used, the blade area normally appears in the analysis. Hence the expression for the thrust coefficient, when using blade element theory, will contain the ratio of blade area to disc area. This ratio is known as the rotor solidity and is denoted by s (or σ).

The definition of solidity is given by:

$$s = \frac{NcR}{\pi R^2} = \frac{Nc}{\pi R} \tag{3.16}$$

which leads to:

$$dC_T = sC_L x^2 \, dx \tag{3.17}$$

Integrating along the blade span gives the rotor thrust coefficient as:

$$C_T = s \int_0^1 C_L x^2 \, dx \tag{3.18}$$

The torque for a single blade, when non-dimensionalised, becomes (noting that there is a change to the integration limits with the use of x as the integrating variable):

$$dC_{Q_1} = \frac{c}{\pi R} (\phi C_L + C_D) x^3 \, dx \tag{3.19}$$

and for N blades of constant chord:

$$dC_Q = s(\phi C_L + C_D) x^3 \, dx \tag{3.20}$$

Integrating along the span gives the rotor torque coefficient as:

$$C_Q = s \int_0^1 (\phi C_L + C_D) x^3 \, dx$$
$$= s \int_0^1 (\lambda C_L x^2 + C_D x^3) dx \tag{3.21}$$

The rotor power requirement is given by:

$$P = \Omega Q \tag{3.22}$$

Note that the power coefficient is defined as:

$$C_P = \frac{P}{\frac{1}{2}\rho(\Omega R)^2 \cdot A \cdot \Omega R}$$
$$= \frac{P}{\frac{1}{2}\rho(\Omega R)^3 \cdot A}$$
$$= \frac{\Omega Q}{\frac{1}{2}\rho(\Omega R)^3 \cdot A}$$
$$= \frac{Q}{\frac{1}{2}\rho(\Omega R)^2 \cdot A \cdot R} = C_Q \tag{3.23}$$

Hence C_P and C_Q are identical in value (essentially there is a Ω term multiplying both the numerator and denominator).

The definitions used in this book contain a half in the denominator. This is not universal as some analyses omit the half. It is also apparent that the normalizing factors are based, as in the lift coefficient, on a dynamic pressure and a reference area. The thrust, torque and power coefficients use the disc area. However, the introduction of blade element theory (BET) gives equations which feature the blade area, so another set of coefficients can be defined to reflect this:

$$\frac{T}{\frac{1}{2}\rho(\Omega R)^2 \cdot NcR} = \frac{T}{\frac{1}{2}\rho(\Omega R)^2 \cdot A \cdot (NcR/A)} = \frac{C_T}{s}$$

$$\frac{Q}{\frac{1}{2}\rho(\Omega R)^2 \cdot R \cdot NcR} = \frac{Q}{\frac{1}{2}\rho(\Omega R)^2 \cdot R \cdot A \cdot (NcR/A)} = \frac{C_Q}{s} \qquad (3.24)$$

$$\frac{P}{\frac{1}{2}\rho(\Omega R)^3 \cdot NcR} = \frac{P}{\frac{1}{2}\rho(\Omega R)^3 \cdot A \cdot (NcR/A)} = \frac{C_P}{s}$$

These are related to the original coefficients via the rotor solidity.

To evaluate Equations 3.18 and 3.21 it is necessary to know the spanwise variation of blade incidence α and to have blade section data which give C_L and C_D as functions of α. The equations can then be integrated numerically. Since α is given by $(\theta - \phi)$, its distribution depends upon the variations of θ, the blade pitch, and $(V_C + V_i)$, the total induced velocity, represented by the inflow factor λ. Useful approximations can be made, however, which allow analytical solutions with, in most cases, only small loss of accuracy.

3.2   Thrust Approximations

If the blade incidence α is measured from the no-lift line and stall and compressibility effects can be neglected, the section lift coefficient can be approximated by the linear relation:

$$C_L = a\alpha = a(\theta - \phi) \qquad (3.25)$$

where the two-dimensional lift slope factor a has a value of about 5.7. Analyses of potential flow by Glauert [1] give a value for the lift curve slope of 2π per radian. The value of 5.7 is more representative allowing for losses due to viscous effects. This value also has the added benefit of being equivalent to 0.1 per degree. Equation 3.18 then takes the form:

$$C_T = sa \int_0^1 (\theta - \phi)x^2 \, dx$$

$$= sa \int_0^1 (\theta x^2 - \lambda x) \, dx \qquad (3.26)$$

For a blade of zero twist, θ is constant. For uniform induced velocity – as assumed in simple momentum theory – the inflow factor λ is also constant. In these circumstances Equation 3.26 integrates readily to:

$$C_T = sa\left[\frac{\theta}{3} - \frac{\lambda}{2}\right]$$
(3.27)

Conventionally, modern blades incorporate negative twist, decreasing the pitch angle towards the tip with the objective of evening out the blade loading distribution. Thus θ takes a form such as:

$$\theta = \theta_0 - \kappa x$$
(3.28)

Here, κ is a linear twist expressed over the *entire* blade from rotor centre to blade tip. Modern blades can have twist variations which are nonlinear. Using this form, the thrust coefficient becomes:

$$C_T = sa\left[\frac{\theta}{3} - \frac{\kappa}{4} - \frac{\lambda}{2}\right]$$
(3.29)

The first two terms can be combined by introducing the blade pitch angle at 75% radius:

$$\theta_{75\%} = \theta_0 - \frac{3}{4}\kappa$$
(3.30)

so the thrust coefficient becomes:

$$C_T = sa\left[\frac{\theta_{75\%}}{3} - \frac{\lambda}{2}\right]$$
(3.31)

and the relation in Equation 3.27 is restored.

Thus a blade with linear twist has the same thrust coefficient as one of constant θ equal to that of a linearly twisted blade at three-quarters radius.

Equation 3.27 expresses the rotor thrust coefficient as a function of pitch angle and inflow ratio. For a direct relationship between thrust coefficient and pitch setting, we need to remove the λ term. This can be achieved by recalling the link between thrust and induced velocity provided by the momentum theorem.

For the rotor in hover, this is Equation 2.12, which on incorporation with Equation 3.27 leads to:

$$C_T = sa\left[\frac{\theta}{3} - \frac{1}{2}\left(\frac{1}{2}\sqrt{C_T}\right)\right]$$

$$\theta = 3\left[\frac{C_T}{sa} + \frac{1}{4}\sqrt{C_T}\right]$$
(3.32)

in which for a blade with a linear twist, θ is taken at three-quarters radius. It is readily seen that correspondingly the direct relationship between θ and λ is:

$$\lambda = \frac{sa}{16}\left[\sqrt{1 + \frac{64}{3sa}\theta} - 1\right] \tag{3.33}$$

3.3 Non-uniform Inflow

An assumption has been made so far, which is that the induced velocity is uniform across the rotor disc. This is an idealized situation, which in fixed-wing terms refers to the elliptically loaded wing which, itself, generates a uniform downwash. The effect of non-uniformity can be introduced by dividing the rotor disc into elements of concentric annuli, where we can treat each annulus as a lifting element and apply BET and momentum theory as before. By restricting the analysis to a generic annulus, over which the downwash can be considered constant, the restriction of uniformity of downwash can be removed. We restrict the analysis to hover and use λ_i instead of λ.

If we have an annulus of radius r and width dr, the thrust produced by this annulus can be expressed using BET, giving:

$$
\begin{aligned}
dT &= \frac{1}{2}\rho(\Omega r)^2 \cdot Nc\, dr \cdot a\left(\theta - \frac{V_i}{\Omega r}\right) \\
&= \frac{1}{2}\rho\Omega^2 \cdot Nc\, dr \cdot a\left(\theta r^2 - \lambda_i rR\right)
\end{aligned} \tag{3.34}
$$

This thrust is expressed using momentum theory as:

$$
\begin{aligned}
dT &= \rho \cdot 2\pi r\, dr \cdot V_i \cdot 2V_i \\
&= 4\rho\pi r\, dr V_i^2 \\
&= 4\rho\pi r\, dr(\Omega R)^2 \cdot \lambda_i^2
\end{aligned} \tag{3.35}
$$

Equating gives:

$$\lambda_i^2 + \frac{sa}{8}\lambda_i - \frac{sa}{8}\theta x = 0$$

$$\lambda_i = \frac{-\dfrac{sa}{8} + \sqrt{\left(\dfrac{sa}{8}\right)^2 + \dfrac{sa}{2}\theta x}}{2} \tag{3.36}$$

$$\lambda_i = \frac{sa}{16}\left[\sqrt{1 + \frac{32}{sa}\theta x} - 1\right]$$

that is we have a quadratic equation in λ_i.

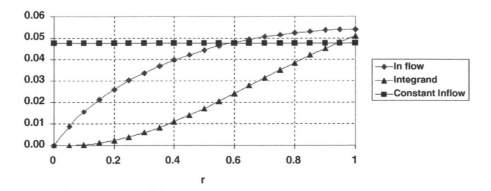

Figure 3.5   Non-uniform inflow: variation of inflow (λ) and integrand ($\theta x^2 - \lambda x$) along blade

The inflow distribution may now be calculated as a function of x and the thrust evaluated from Equation 3.26.

As a numerical example let us consider the case of a blade having linear twist, from a collective pitch setting of 12° to 6° at the tip (the root cutout can be ignored for this purpose). The rotor solidity (s) is 0.08 and the lift curve slope value (a) is 5.7. The value of the pitch angle at 75% radius is then $\theta_{75\%} = 7.5°$.

3.3.1   Constant Downwash

Applying Equation 3.32 for the three-quarters radius point, at which θ is 7.5°, gives a thrust coefficient $C_T = 0.0091$. Turning now to Equation 3.36, the non-uniform λ varies along the span as shown in Figure 3.5.

Superficially this is greatly different from a constant value – Equation 2.12. Nevertheless, on evaluating Equation 3.36 the variation of ($\theta r^2 - \lambda r$) is as shown in the figure, from which the integrated value of thrust coefficient is $C_T = 0.0092$. Thus the assumption of constant inflow has led to underestimating the thrust by a mere 1.7%. The result agrees well with Bramwell's general conclusion (p. 93) and confirms that uniform inflow may be assumed for many, perhaps most, practical purposes.

3.4   Ideal Twist

The relation in Equation 3.36 contains one particular situation when λ becomes a constant, that is if the term θx is itself constant, then:

$$\theta x = \theta_{TIP} \tag{3.37}$$

θ_{TIP} being the pitch angle at the tip. This nonlinear twist is not physically realizable near the root but the case is of interest because, as momentum theory shows, uniform induced velocity corresponds to minimum induced power. The analogy with elliptic loading for a fixed-wing aircraft is again recalled. The twist defined in Equation 3.37 is known as *ideal twist*. Inserting in

Equation 3.26 gives:

$$C_T = sa \int_0^1 \left(\frac{\theta_{TIP}}{x} x^2 - \lambda x \right) dx$$

$$= sa \int_0^1 (\theta_{TIP} - \lambda) x \, dx \qquad (3.38)$$

$$= \frac{sa}{2} (\theta_{TIP} - \lambda)$$

and, since $\lambda = x\phi = \phi_{TIP}$, the tip inflow angle, Equation 3.38 can be expressed as:

$$C_T = \frac{sa}{2} (\theta_{TIP} - \phi_{TIP}) \qquad (3.39)$$

With ideal twist and the resulting constant value of λ we find Equation 3.36 simplifies to:

$$\lambda_i = \frac{sa}{16} \left[\sqrt{1 + \frac{32}{sa} \theta_{TIP}} - 1 \right] \qquad (3.40)$$

and the direct relationship between θ and C_T is now:

$$\theta_{TIP} = 2 \frac{C_T}{sa} + \frac{1}{2} \sqrt{C_T} \qquad (3.41)$$

Some pitch angles for ideal twist and linear twist are compared in Figure 3.6.

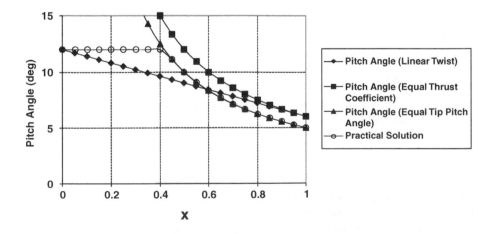

Figure 3.6 Ideal twist and linear twist compared

Figure 3.7 View along Boeing Chinook main rotor blade showing twist

The inboard end of the blade is assumed to be at $r = 0$, ignoring for the purposes of comparison the practical necessity of a root cutout. The linear twist is assumed to vary from 12° pitch at the root to 6° at the tip. Figure 3.7 shows a typical main rotor blade. The built-in twist is readily observed. A straightforward comparison is when the ideal twist has the same pitch at the tip—we see that unrealistically high pitch angles are involved at 40% radius and inboard. A more useful comparison is at equal thrust for the two blades. From Equations 3.32 and 3.41 it follows that for the same thrust coefficient the pitch angle at two-thirds span with the ideal twist is the same as that at three-quarters span with the linear twist, which for the case in point is 7.5°. Thus the ideal twist is given by:

$$\theta x = \theta_{\text{TIP}} = 7.5° \times \frac{2}{3} = 5.0° \tag{3.42}$$

This case is also shown in Figure 3.6. The two twist distributions give the same pitch angle when:

$$x = 1 - \sqrt{\frac{1}{6}} \simeq 0.59 \tag{3.43}$$

Again the ideal twist leads to high pitch angles further inboard but a practical solution, losing little in induced power, might be to transfer to constant pitch from about 40% rotor radius inwards.

3.5 Blade Mean Lift Coefficient

Characteristics of a rotor obviously depend on the lift coefficient at which the blades are operating and it is useful to have a simple approximate indication of this. The blade mean lift coefficient provides such an indication. As the name implies, the mean lift is that which, applied uniformly along the blade span, would give the same total thrust as the actual blade. Writing the mean lift coefficient as $\overline{C}_L$ we have, from Equation 3.18:

$$C_T = s \int_0^1 C_L x^2 \, dx$$

$$= s\overline{C}_L \int_0^1 x^2 \, dx \tag{3.44}$$

$$= \frac{1}{3} s\overline{C}_L$$

from which:

$$\overline{C}_L = \frac{3C_T}{s} \tag{3.45}$$

The parameter C_T/s has been previously discussed and this gives another reason for the preference some workers have for using it as the definition of thrust coefficient instead of C_T.

Blades usually operate in the C_L range 0.3–0.6, so typical values of C_T/s are between 0.1 and 0.2. Typical values of C_T are an order of 10 smaller as solidity values are in the region of 0.1.

3.6   Power Approximations

From Equation 3.20 the differential power coefficient dC_P $(= dC_Q)$ may be written as:

$$dC_P = dC_Q$$

$$= s(\phi C_L + C_D)x^3 \, dx$$

$$= sC_L \phi x^3 \, dx + sC_D x^3 \, dx \tag{3.46}$$

$$= sC_L \lambda x^2 \, dx + sC_D x^3 \, dx$$

$$= dC_{Pi} + dC_{P0}$$

where dC_{Pi} is the differential power coefficient associated with induced flow and dC_{P0} is that associated with blade section profile drag. The first term, using Equation 3.17, is simply:

$$dC_{Pi} = \lambda \, dC_T \tag{3.47}$$

Thus:

$$dC_P = \lambda \, dC_T + sC_D x^3 \, dx \tag{3.48}$$

whence:

$$C_P = \int_{x=0}^{x=1} \lambda \, dC_T + \int_0^1 sC_D x^3 \, dx \tag{3.49}$$

Assuming uniform inflow and a constant profile drag coefficient C_{D0}, we have the approximation:

$$C_P = \lambda \cdot C_T + \frac{sC_{D0}}{4}$$ (3.50)

In the hover, where:

$$\lambda = \frac{1}{2}\sqrt{C_T}$$ (3.51)

this becomes:

$$C_P = \frac{1}{2}(C_T)^{3/2} + \frac{sC_{D0}}{4}$$ (3.52)

The first term of Equations 3.50 or 3.52 agrees with the result from simple momentum theory (Equation 2.12). The present λ, defined by Equation 3.14, includes the inflow from climbing speed V_C (if any), so the power coefficient term includes the climb power:

$$P_{CLIMB} = V_C \cdot T$$ (3.53)

The total induced power in hover or climbing flight is generally two or three times as large as the profile power. The chief deficiency of the formula in Equation 3.50 in practice arises from the assumption of uniform inflow. Bramwell (p. 94ff.) shows that for a linear variation of inflow the induced power is increased by approximately 13%. This and other smaller correction factors such as tip loss (Section 3.7) are commonly allowed for by applying an empirical factor k_i to the first term of Equation 3.50, so that as a practical formula:

$$C_P = k_i \cdot \lambda \cdot C_T + \frac{sC_{D0}}{4}$$ (3.54)

is used, in which a suggested value of k_i is 1.15. The combination of Equations 3.54 and 3.27 provides adequate accuracy for many performance problems.

For the hover, we have:

$$C_P = k_i \cdot \frac{1}{2}(C_T)^{3/2} + \frac{sC_{D0}}{4}$$ (3.55)

The figure of merit M may be written:

$$M = \frac{C_{P\,IDEAL}}{C_{P\,ACTUAL}} = \frac{(C_T)^{3/2}}{k_i \cdot (C_T)^{3/2} + \frac{sC_{D0}}{2}}$$ (3.56)

which demonstrates that for a given thrust coefficient a high figure of merit requires a low value of the product sC_{D0}. Using a low solidity seems an obvious way to this end but it must be tempered because the lower the solidity, the lower the blade area, which means the higher the blade angles of incidence required to produce the thrust and the profile drag may then be increased significantly from either Mach number effects or the approach of stall. A low solidity, subject to retaining a good margin of incidence below the stall, would appear to be the formula for producing an efficient design.

For accurate performance work the basic relationships in Equations 3.18 and 3.21 are integrated numerically along the span. Appropriate aerofoil section data can then be used, including both compressibility effects and stalling characteristics. Further reference to numerical methods is made in Chapter 6.

3.7   Tip Loss

A characteristic of the actuator disc concept is that the linear theory of lift is maintained to the perimeter of the disc. Physically, as described in Chapter 2, we suppose the induced velocity, in which the pressure is above that of the surrounding air, to be contained entirely below the disc in a well-defined streamtube surrounded by air at rest relative to it. In reality, because the rotor consists of a finite number of separate blades, some air is able to escape outwards between the tips, drawn out by the tip vortices. Thus the total induced flow is less than the actuator disc theory would prescribe, so that for a given pitch setting of the blades the thrust is somewhat lower than that given by Equation 3.27. The deficiency is known as tip loss and is shown by a rapid falling off of lift over the last few per cent of span near the tip, in a typical blade loading distribution such as that of Figure 2.28.

Although several workers have suggested approximations [Bramwell (p. 111) quotes Prandtl, Johnson (p. 60) quotes in addition Sissingh and Wheatley], no exact theory of tip loss is available. A common method of arriving at a formula is to assume that outboard of a station $r = BR$ the blade sections produce drag but no lift. Then the thrust integral in Equation 3.26 is replaced by (which is a change in the upper limit of the thrust integral, effectively ignoring the outer tip region of the blade):

$$C_T = sa \int_0^B \left(\theta x^2 - \lambda x \right) \, dx \tag{3.57}$$

whence is obtained, for uniform inflow and zero twist:

$$C_T = sa \left(\frac{\theta B^3}{3} - \frac{\lambda B^2}{2} \right) \tag{3.58}$$

With a typical value $B = 0.97$ or 0.98 Equation 3.58 yields between 5% and 10% lower thrust than Equation 3.27 for a given value of θ.

To obtain the effect on rotor power at a given thrust coefficient, we need to express the increase in induced velocity corresponding to the effective reduction of disc area. Since the latter is affected by a factor B^2 and the induced velocity is proportional to the square root of disc loading (Equation 2.8), the increase in induced velocity is by a factor $1/B$. The rotor induced

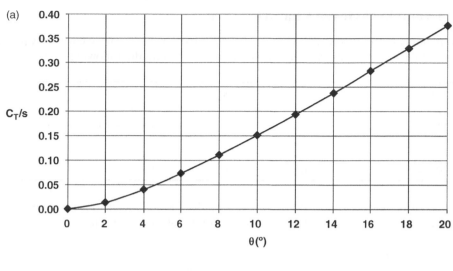

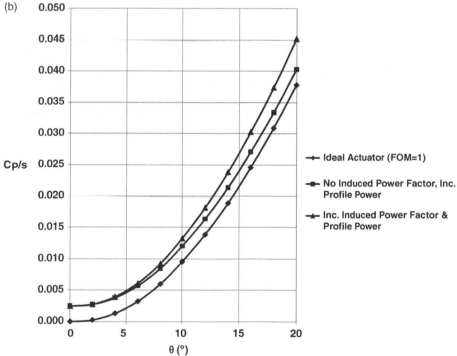

Figure 3.8 (a) Thrust coefficient v pitch angle. (b) Power coefficient v pitch angle.

power in hover thus becomes:

$$C_{\mathrm{Pi}} = \frac{1}{B} \cdot \frac{(C_{\mathrm{T}})^{3/2}}{2} \qquad (3.59)$$

Typically this amounts to 2–3% increase in induced power. The factor can be incorporated in the overall value assumed for the empirical constant k in Equation 3.55.

3.8   Example of Hover Characteristics

Corresponding to C_L/α and C_D/C_L characteristics for fixed wings, we have C_T/θ and C_P/C_T for the helicopter in hover. An example has been evaluated using the following data:

Blade radius	R	6 m
Blade chord (constant)	c	0.5 m
Blade twist	κ	Linear from 12° at root to 6° at tip
Number of blades	N	4
Empirical constant	k_i	1.13
Blade profile drag coefficient (constant)	C_{D0}	0.010

The variation of C_T/s with θ is shown in Figure 3.8a. The nonlinearity results from the $\sqrt{C_T}$ term in Equation 3.32. The variation of C_P/s with θ is calculated for three cases:

- $k = 1.13$, Equation 3.55;
- $k = 1.0$, Equation 3.50, the simple momentum theory result;
- figure of merit $M = 1.0$, which assumes $k_i = 1.0$ and $C_{D0} = 0$.

Over the range shown (Figure 3.8b), using the factor $k = 1.13$ results in a power coefficient 0–9% higher than that obtained using simple momentum theory. The curve for $M = 1$ is of course unrealistic but gives an indication of the division of power between induced and profile components.

(Rotor performance characteristics are sometimes plotted as C_P/s versus C_T/s. This type of plot is known as a hover polar.)

Reference

1. Glauert, H. (1983) *The Elements of Aerofoil and Airscrew Theory*, Reissued in the Cambridge Science Classics Series, Cambridge University Press.

4

Rotor Mechanisms for Forward Flight[1]

4.1 The Edgewise Rotor

In level forward flight the rotor is essentially edgewise on to the air stream, a basically unnatural state for propeller functioning. This is shown in Figure 4.1. Practical complications which arise from this have been resolved by the introduction of mechanical devices, the functioning of which in turn adds to the complexity of the aerodynamics.

Figure 4.2 pictures the rotor disc as seen from above. Blade rotation is in an anticlockwise sense with rotational speed Ω. Forward flight velocity is V and the ratio $V/\Omega R$, R being the blade radius, is known as the advance ratio and given the symbol μ. It has a value normally within the range 0.0 to 0.5. Azimuth angle ψ is measured from the downstream blade position: the range $\psi = 0°-180°$ defines the advancing side and that from $180°-360°$ (or $0°$) the retreating side.

A blade is shown in Figure 4.2 at 90° and again at 270°. These are the positions of maximum and minimum relative air velocity normal to the blade, the velocities at the tip being $(\Omega R + V)$ and $(\Omega R - V)$, respectively. If the blade were to rotate at fixed incidence, then owing to this velocity differential, much more lift would be generated on the advancing side than on the retreating side. Calculated pressure contours for a fixed-incidence rotation with $\mu = 0.3$ are shown in Figure 4.3. For this situation, about four-fifths of the total lift is produced on the advancing side. The consequences of this imbalance would be large oscillatory bending stresses at the blade roots and a large rolling moment on the vehicle tending to roll the aircraft towards the retreating side. Both structurally and dynamically the helicopter would be unflyable.

Clearly a cyclical variation in blade incidence is needed to balance lift on the two sides. If we permit the blade pitch to vary sinusoidally as the blade rotates around the azimuth to an amount which balances out the rolling moment of the rotor, the contours of pressure level for this roll-balanced lift distribution are of the type shown in Figure 4.4.

The mean pressure level is now lower, the lift on the advancing side being greatly reduced, with only small compensation on the retreating side. The fore and aft sectors now carry the main lift load. The total lift can be restored in some degree by applying a general increase in blade

---

[1] Portions of this Chapter have been taken from *The Foundations of Helicopter Flight* by Simon Newman, Elsevier, 1994.

Basic Helicopter Aerodynamics, Third Edition. John Seddon and Simon Newman.
© 2011 John Wiley & Sons, Ltd. Published 2011 by John Wiley & Sons, Ltd.

Figure 4.1   Main rotor alignment in forward flight

incidence level through the pilot's control system (Section 4.3), but as this is performed the retreating blade, which is producing lift at relatively low airspeed, must ultimately stall. In addition, compressibility effects such as shock-induced flow separation must be considered, both on the advancing side where the Mach number is highest and on the retreating side where lower Mach number is combined with high blade incidence. Since the degree of load asymmetry across the disc increases with forward speed, the retreating-blade stall and its associated effects determine the maximum possible flight speed of the vehicle. For the conventional helicopter a speed of about 400 km/h (250 mph) is usually regarded as the upper limit.

While this technique would balance out the overall rolling moment, the rigid connection of the blades to the hub would cause a significant amount of vibratory forcing to be transmitted to the rotor hub and hence to the helicopter. The mechanisms which can avoid this are now described. The widely adopted method of achieving this is by use of flapping hinges, first introduced by Juan de la Cierva around 1923. The blade is freely hinged as close as possible to

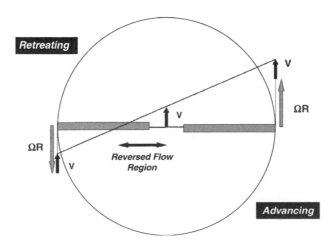

Figure 4.2   Velocity components

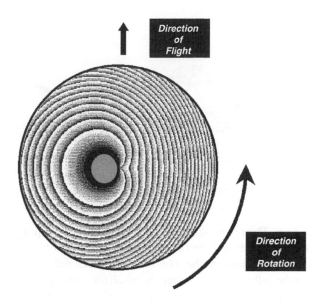

Figure 4.3   Pressure contours without roll trim

the root, allowing it to flap up and down out of the plane of rotation as it rotates about the rotor shaft. The presence of free hinges means that blade root stresses are avoided and no rolling moment is communicated to the airframe. The blades are now under the influence of two types of moments. The lift force is trying to flap the blades upwards by creating a moment about the flapping hinge – moving the blades out of the rotation plane – while the centrifugal force is working in opposition – driving the blades back towards the rotation plane. In hover, the blade position is stable and these two moments cancel out. Figure 4.5 shows a Sikorsky S61NM

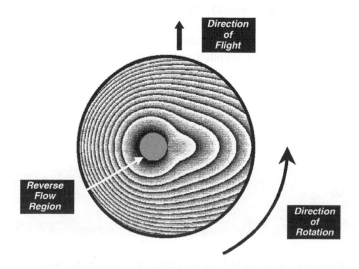

Figure 4.4   Pressure contours with roll trim

Figure 4.5   Sikorsky S61NM helicopter approaching touchdown (coned rotor)

hovering prior to landing. The rotor blades are 'coned up' as the flapping hinges relieve the flapping moment of the lift loads and the two moments are in equilibrium. Figure 4.6 shows the aircraft after touching down where the rotor thrust has been reduced to zero and the rotor disc is now flat, under the influence of the centrifugal effects only (*gravity has a small effect which is usually neglected – the main rotor tip experiences centripetal accelerations of 750–1000 g*).

As the rotor moves into horizontal flight the situation of Figure 4.2 establishes itself. The incident flow over the blades will depend on their azimuthal position. The blade on the advancing side experiences an increased incident flow and the lift will increase, overcoming the centrifugal moment, so the blade will now flap upwards. As it flaps upwards, a downward flow is superimposed on the blade and the lift will reduce. See Figure 4.7. This continues until the blade flaps to its highest position. The reverse effect occurs for the blade on the retreating side where the blade will flap downwards to its lowest position. The maximum velocity change is at the 90° and 270° azimuth. It will be shown that the extreme blade flapping angles are achieved very close to 90° of blade rotation around the azimuth. The result is that the blades will flap up at the front of the rotor and down at the rear and the disc will tilt rearwards. It is normally accepted that the thrust force is aligned with the normal to the rotor disc. We have the situation where the rotor thrust is now inclined rearwards and forward motion is not possible. In order for the rotor to supply forward propulsion, the rotor disc must be tilted forwards, directly opposing the natural effect of blade flapping. This can only be achieved by altering the blade lift through a pitch change. This is known as cyclic pitch and will be discussed later.

Figure 4.6   Sikorsky S61NM helicopter after touchdown (flat rotor)

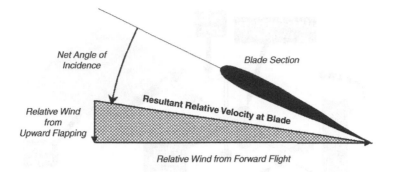

Net Angle of Incidence

Blade Section

Resultant Relative Velocity at Blade

Relative Wind from Upward Flapping

Relative Wind from Forward Flight

Figure 4.7   Incident velocity components

This pitch angle movement is provided by a pitch bearing, known alternatively as the feathering hinge, linked to a control system operated by the pilot (Section 4.3).

An additional feature of the asymmetry in velocity across the disc is that there exists a region on the retreating side where the flow over the blade is actually reversed. At 270° azimuth the resultant velocity at a point r of span is:

$$U = \Omega r - V \tag{4.1}$$

or non-dimensionally:

$$u = \frac{U}{\Omega R} = x - \mu \tag{4.2}$$

Thus the flow over the blade is reversed inboard of the point $x = \mu$. It will be apparent that the reversed flow boundary is a circle of diameter μ, centred at $x = \mu/2$ on the 270° azimuth. Dynamic pressure in this region is low, so the effect of the reversed flow on the blade lift is small, usually negligible from a performance aspect for advance ratios up to 0.4. Very precise calculations may require the reversed flow region to be taken into account and it may be important also in studies of blade vibration. Some advanced rotor concepts have required the rotor rotational speed to be reduced, even halted. This will cause the reversed flow region to become a significant part of the rotor disc.

A flapping blade in rotation sets up Coriolis moments in the plane of the disc, and to relieve this it is usual to provide a second hinge, the lead–lag hinge, normal to the disc plane, allowing free in-plane motion. This may need to be fitted with a mechanical damper to ensure dynamic stability.

The standard articulated blade thus possesses this triple movement system of flapping hinge, lead–lag hinge and pitch bearing in a suitable mechanical arrangement, located inboard of the lifting blade itself. The principles are illustrated in Figure 4.8.

Kaman aircraft use a slightly different system where blade pitch is controlled by a trailing edge servo flap. This is deflected by the pilot's controls, which generates a moment causing the blade to elastically bend in pitch. These servo flaps can be seen in Figure 4.9 which shows the KMax while Figure 4.10 shows the Seasprite.

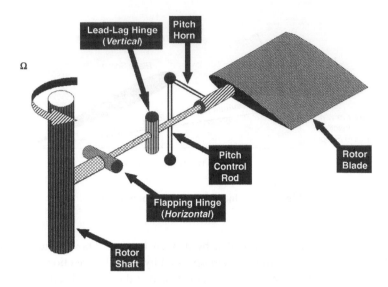

Figure 4.8 Principles of an articulated rotor hinge system

Figure 4.9 Kaman KMax (Courtesy Stewart Penney)

Figure 4.10 Kaman Seasprite (Courtesy US Navy)

Strictly the blade root bending stress and helicopter rolling moment are eliminated by flapping only if the hinge is located on the axis of rotation. This is impracticable for a rotor with more than two blades, so residual moments do exist. These are not important, however, if the offset of the hinge from the axis is only a few per cent of blade radius – which is normal. The flapping hinge is therefore normally made the innermost, with an offset of 3–4%. The lag hinge and pitch bearing can be more freely disposed: sometimes the former is the farther out of the two. The order of the hinges has a significant effect on the blade dynamic behaviour and if the pitch bearing is placed inboard of either or both of the flap and lag hinges, kinematic couplings can be generated.

The total mechanical complexity of an articulated rotor is substantial. Hinge bearings operate under high centrifugal loads, so service and maintenance requirements are severe. Hinges, dampers and control rods make up a bulky rotor head, which is likely to have a high parasitic drag – perhaps as much as the rest of the helicopter.

In modern rotors, the flapping and lag hinges are often replaced by flexible elements which allow the flapping and lead–lag motions of the blades to take place, albeit with a degree of stiffness not present with free hinges. With such hingeless rotors, bending stresses and rolling moments reappear, in moderation only but sufficient to modify the stability and control characteristics of the helicopter (Chapter 8). The effect of a flexible flapping element can usually be calculated by equating it to a hinged blade with larger offset (10–15%). The use of a hingeless rotor is one way of reducing the parasitic drag of the rotor head. A pitch bearing mechanism is of course needed for rotor control, as with the articulated rotor. The hingeless rotor of the Agusta Westland Lynx helicopter is pictured in Figure 4.11.

4.2 Flapping Motion

To examine the flapping motion more fully we assume, unless otherwise stated, that the flapping hinge is on the axis of rotation. This simplifies the considerations without hiding anything of significance.

Referring to Figure 4.12, the flapping takes place under conditions of dynamic equilibrium, about the hinge, between the aerodynamic lift (the external forcing function), the centrifugal

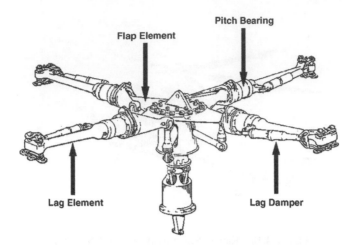

Figure 4.11   Main rotor hub of Agusta Westland Lynx helicopter (Courtesy Agusta Westland)

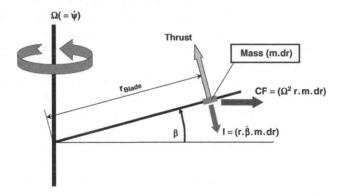

Figure 4.12   Blade forces in flapping

force (the 'spring' or restraining force) and the blade inertia. (The aerodynamic lift varies as the blade flapping responds in a manner which acts as a damper.) In other words, the once-per-cycle oscillatory motion is that of a dynamic system in resonance. The flapping moment equation is seen to be:

$$\int_0^R mr^2\ddot{\beta}\ \mathrm{d}r = \int_0^R r\ \mathrm{d}T - \int_0^R m\beta r^2\Omega^2\ \mathrm{d}r \tag{4.3}$$

We shall return to this equation later.

The centrifugal force supplies, by far, the largest force acting on the blade and it creates the moment which provides an essential stability to the flapping motion – essentially it acts as a spring. The degree of stability is highest in the hover condition (where the flapping angle is constant) and decreases as the advance ratio increases. Bramwell's consideration of the flapping equation (p. 153ff.) leads in effect to the conclusion that the motion is dynamically stable for all realistic values of μ.

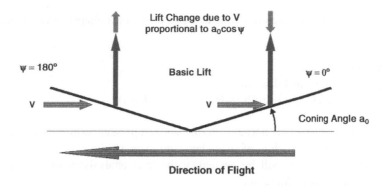

Figure 4.13 Longitudinal lift asymmetry which leads to lateral tilt

For normal forward flight, the maximum flapping velocities $(\dot{\beta})$ occur where the resultant air velocity is at its highest and lowest, that is at 90° and 270° azimuth. Maximum displacements occur 90° later, that is at 180° (upwards) and 0° (downwards). As will be seen, the flapping has a natural frequency very close to that of the rotor speed. It is therefore close to a resonant condition which will give a phase delay of near 90° (exactly 90° for a zero hinge offset). While it is near resonance, the aerodynamics provides the damping to keep the responses under control. As already outlined, these displacements mean that the plane of rotation of the blade tips, the tip-path plane (TPP), is tilted backwards relative to the plane normal to the rotor shaft, the shaft normal plane (SNP).

In hover, the blades cone upwards at a constant angle to the SNP, known as the coning angle, usually denoted by a_0. Its existence has an additional effect on the orientation of the TPP during rotation in forward flight.

Figure 4.13 shows that, because of the coning angle, the flight velocity V has a lift-increasing effect on a blade at 180° (the forward blade) and a lift-decreasing effect on a blade at 0° (the rearward blade). This asymmetry in lift is, we see, at 90° to the side-to-side asymmetry discussed earlier: its effect is to tilt the TPP laterally and since the point of lowest tilt follows 90° behind the point of lowest lift, the TPP is tilted downwards on the advancing side which, with, a rotor rotating anticlockwise from above, is to the right. The coning and disc tilt angles are normally no more than a few degrees.

Since in any steady state of the rotor the flapping motion is periodic, the flapping angle can be expressed in the form of a Fourier series:

$$\beta = a_0 - a_1 \cos \psi - b_1 \sin \psi - a_2 \cos 2\psi - b_2 \sin 2\psi \ldots \tag{4.4}$$

Textbooks vary both in the symbols used and in the sign convention adopted. The use of negative signs for the harmonic terms is a throwback to the emergence of the autogyro where a rearward disc tilt was the norm, and where the coefficients a_1 and b_1 have positive values. For most purposes the series can be limited to the constant and first harmonic terms – which represent the coned rotor and the disc tilt – thus:

$$\beta = a_0 - a_1 \cos \psi - b_1 \sin \psi \tag{4.5}$$

This form will be used in the aerodynamic analysis of the next chapter. For the moment we note that:

- a_0 is the coning angle;
- a_1 is the angle of backward disc tilt;
- b_1 is the angle of sideways disc tilt (*advancing side down*).

The inclusion of second or higher harmonic terms would represent perturbations about the TPP (in-plane weaving) but any such is of secondary importance only.

Timewise derivatives of β will be needed in the later analysis: using the fact that the rotational speed Ω is $d\psi/dt$, these are:

$$\dot{\beta} = \Omega \frac{d\beta}{d\psi} = \Omega(a_1 \sin\psi - b_1 \cos\psi) \tag{4.6}$$

$$\ddot{\beta} = \Omega^2 \frac{d^2\beta}{d\psi^2} = \Omega^2(a_1 \cos\psi + b_1 \sin\psi) \tag{4.7}$$

The transformation to β as a function of ψ, the blade azimuth, makes the solutions more informative and therefore useful. It also has the benefit of causing many scaling terms to cancel out.

4.3   Rotor Control

Control of the helicopter in flight involves changing the magnitude of rotor thrust or its line of action or both. Almost the whole of the control task falls to the lot of the main rotor and it is on this that we concentrate. (The main rotor controls heave, surge/pitch, sway/roll, while yaw is controlled by a tail rotor or similar installation.) A change in line of action of the thrust would, in principle, be obtained by tilting the rotor shaft, or at least the hub, relative to the fuselage. Since the rotor is engine driven (unlike that of an autogyro), tilting the shaft is impracticable. It was attempted with the Cierva W9 as described in Chapter 1. Tilting the hub is possible with some designs but the large mechanical forces required restrict this method to very small helicopters. Use of the feathering mechanism, however, by which the pitch angle of the blades is varied, either collectively or cyclically, effectively transfers to the aerodynamic forces the work involved in changing the magnitude and direction of the rotor thrust.

Blade feathering, or pitch change, could be achieved in various ways. Thus Saunders [1] lists the use of aerodynamic servo tabs, auxiliary rotors, fluidically controlled jet flaps, or pitch links from a control gyro as possible methods.

The widely adopted method, however, is through a swashplate system, illustrated in Figure 4.14 (NFP is the No-Feathering Plane which is discussed later in the chapter.), which shows the operation with collective pitch, while Figure 4.15 shows the operation with cyclic pitch. Carried on the rotor shaft, this embodies two parallel star-shaped plates, the lower of which does not rotate (*swashplate or non-rotating star*) with the shaft but can be tilted in any direction by operation of the pilot's cyclic control column and raised or lowered by means of the collective lever. The upper plate (*spider or rotating star*) is connected by control rods to the feathering hinge mechanisms of the blades and rotates with the shaft, while being constrained to remain parallel to the lower plate through a common bearing assembly. Raising the collective lever thus increases the pitch angle of the blades by the same amount all round (Figure 4.14), while tilting

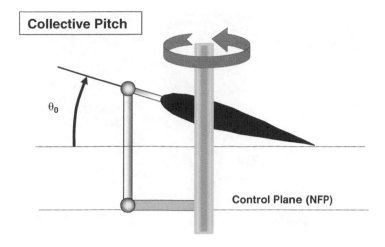

Figure 4.14   Principles of the swashplate system (collective pitch)

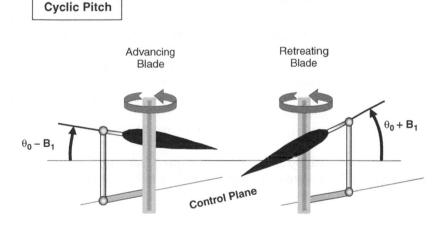

Figure 4.15   Principles of the swashplate system (cyclic pitch)

the cyclic column applies a tilt to the plates and thence a cyclic pitch change to the blades (Figure 4.15), these being constrained to remain at constant pitch relative to the upper plate.

Figure 4.16 shows a typical control layout of a helicopter cockpit. It is an early design and is very simple. The aircraft is a Saunders Roe Skeeter, and while dual controls are provided, the pilot normally sits in the right-hand seat. More modern helicopters look more complicated with other system controls placed on the collective lever or cyclic stick, but are the same in essence.

An increase of collective pitch at constant engine speed increases the rotor thrust (short of stalling the blades), as for take-off and vertical control generally. A cyclic pitch change alters the line of action of the thrust, since the TPP of the blades, to which the thrust is effectively perpendicular, tilts in the direction of the swashplate angle.

Rotor head designs vary considerably in detail as shown in Figures 4.17a–k. Figure 4.17a shows the simplest rotor head arrangement, namely the two-bladed teetering type. Figure 4.17b

Figure 4.16   Pilot's controls of a Saunders Roe Skeeter

shows a typical articulated rotor as used on the Sikorsky S61N. Figure 4.17c also shows a modern rotor of the fully articulated type. It is of an Agusta Westland Merlin and shows its location on top of the fuselage. Items discernible in the figure include the flap and lag elastomeric hinges, the feathering housing, the lag damper, the pitch control rods and the dual load path blade attachment.

Figure 4.17d shows a close-up view of a Merlin rotor head blade attachment and the blade restraint mechanism of the droop and anti-flap stops can be seen as a horizontal pin fitting into a slot. The main rotor blade folding mechanism is also shown. A good impression is gained of the mechanical complexity of this rotor head installation.

Figure 4.17e shows the semi-rigid rotor of the Westland WG30 (with vibration absorber fitted).

Figure 4.17f shows the MBB Bo 105 rotor head with pendulum vibration absorbers fitted to each blade attachment. The modern strap construction of the Bell 412 is shown in Figure 4.17g, which also has pendulum vibration absorbers installed.

Figure 4.17h shows the 'Starflex' rotor of the Eurocopter Ecureuil.

Figure 4.17i shows the complexity of a coaxial rotor system. The aircraft is the Kamov Helix and has two contra-rotating three-bladed rotors on the same shaft.

In comparison, the tail rotor of a Sikorsky S61NM is shown in Figure 4.17j while that of a Merlin is shown in Figure 4.17k.

Figure 4.18 is an interior view of the cockpit of a modern helicopter (dual controls).

The collective pitch lever is down at seat level on the pilot's left (right-hand seat); the cyclic control stick is directly in front between the knees. The foot pedals control the collective pitch of the tail rotor (normally its only control), the purpose of which is to balance the torque of the main rotor, or when required to change the heading of the aircraft.

Cyclic pitch on the main rotor implies a blade angle changing with azimuth, relative to the SNP. The once-per-cycle periodicity means that the pitch angle can be described mathematically by a negative Fourier series, in like manner to that used for the flapping angle.

We write:

$$\theta = \theta_0 - A_1 \cos \psi - B_1 \sin \psi - A_2 \cos 2\psi - B_2 \sin 2\psi \ldots \qquad (4.8)$$

Figure 4.17 (a) Main rotor head of a Bell JetRanger helicopter. (b) Main rotor head of a Sikorsky S61N helicopter. (c) Agusta Westland Merlin main rotor head. (d) Agusta Westland Merlin main rotor head. (e) Semi-rigid main rotor head of a Westland WG30 helicopter. (f) Main rotor head of a Bolkow Bo105 helicopter. (g) Main rotor head of a Bell 412 helicopter. (h) Starflex main rotor head of a Eurocopter Ecureuil helicopter. (i) Coaxial rotor assembly for a Kamov Helix helicopter (Courtesy US Navy). (j) Tail rotor of a Sikorsky S61NM helicopter. (k) Tail rotor of an Agusta Westland Merlin helicopter. The mechanical connections between the rotor controls and blades differ between main and tail rotors. This is because they operate under different aerodynamic requirements. With a main rotor, the connections are aligned so that blade flapping does not influence the blade pitch angle. With a tail rotor, the control geometry is differently aligned to give a coupling between the blade flapping and pitching. This is sometimes referred to as a delta 3 hinge.

Figure 4.17   (*Continued*)

The physical design of a swashplate system only permits once per revolution changes in the pitch angle so only the constant and first harmonic terms are normally required:

$$\theta = \theta_0 - A_1 \cos \psi - B_1 \sin \psi \tag{4.9}$$

- The constant term θ_0 represents the collective pitch.
- The terms in ψ represent the cyclic pitch:
  - The factor A_1, which applies maximum pitch when the blades are at $0°$ and $180°$, is referred to as the lateral cyclic coefficient because the rotor response, phased $90°$, produces a control effect in the lateral sense.
  - The factor B_1 is the longitudinal cyclic coefficient.

The value of pitch angle would be different if a different reference plane were used. In any flight condition, there is always one plane relative to which the blade pitch remains constant with azimuth. This, by definition, is the plane of the swashplate, which is therefore known as the control plane or, referring to the elimination of cyclic pitch variation, the no-feathering plane

Figure 4.18 Cockpit of an Agusta Westland Merlin helicopter

(NFP). NFP, though not fixed in the aircraft, is a useful adjustable datum for the measurement of aerodynamic characteristics considered in the next chapter.

In some contexts it is useful to refer to the axes TPA and NFA, perpendicular to the TPP and NFP, rather than to the planes themselves. Generally in forward flight these two axes and also the shaft axis will be away from the vertical (i.e. the normal to the flight path). Figure 4.19

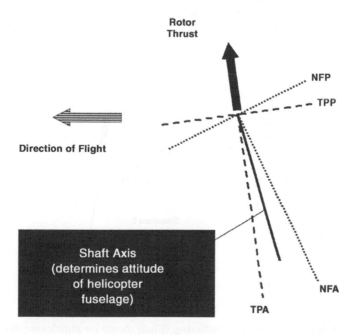

Figure 4.19 A possible juxtaposition of axes in forward flight

shows a common arrangement. The thrust line being inclined in the direction of flight, the TPP normal to it is tilted down at the nose relative to the horizontal (the flight direction). The TPA, being also the thrust line, is away from the vertical as shown. The shaft axis is tilted further from the vertical, the angle with the TPA being the tilt-back angle of the flapping motion. The inclination of the shaft axis to the NFA depends upon the degree of feathering in the helicopter motion.

4.4   Equivalence of Flapping and Feathering

The performance of the rotor blade depends upon its angle of incidence to the TPP. A given blade incidence can be obtained with different combinations of flapping and feathering. Consider the two situations illustrated in Figure 4.20.

These are views from the left side with the helicopter in forward flight in the direction shown. In situation 1 the shaft axis coincides with the TPA; there is therefore no flapping but the necessary blade incidences are obtained from feathering according to Equation 4.9. Blade attitudes at the four quarter points of a rotation are as indicated in the diagram. In situation 2 the shaft axis coincides with the NFA. By definition this means that feathering is zero; the blade angles, however, are obtained from flapping according to Equation 4.5. It is seen that if the feathering and flapping coefficients B_1 and a_1 are equal, the blade attitudes to the TPP are identical around the azimuth in the two situations. The blade perceives a change in nose-down feathering, via the swashplate, as being equivalent to the same angle change in nose-up flapping.

A pilot uses this equivalence in flying the helicopter, for example to trim the vehicle for different positions of the centre of gravity (CG). The rotor thrust, in direction and magnitude, depends upon the inclination of the TPP in space and the incidence of the blades relative to it. The same blade incidence can be achieved, as we have seen, either with nose-up flapping or with the same degree of nose-down feathering, or of course with a combination of the two. By adjusting the relationship, using the cyclic control stick, the pilot is able to

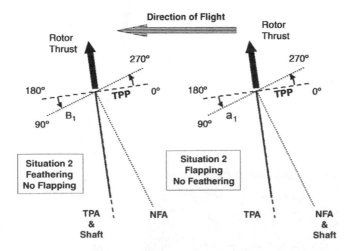

Figure 4.20   Equivalence of flapping and feathering. (Blade chordwise attitudes are shown in the plane of the diagram for azimuth angles of 90° and 270° and normal to the diagram for 0° and 180°)

compensate for different nose-up or nose-down moments in the helicopter, arising from different CG positions. The angle of the shaft axis to the vertical, hence the attitude of the helicopter in space, varies with the CG position but the TPP remains at a constant inclination to the direction of flight.

4.4.1 Blade Sailing

As previously discussed, the flapping behaviour of a rotor blade is governed principally by the balance between the aerodynamic lift and the centrifugal force moments about the flapping hinge. These are both dependent on the square of the rotor speed and so, under atmospherically still conditions, the balance is preserved at any rotor speed. However, when operating in high wind conditions, such as on a ship, the balance can be upset. During any helicopter sortie, the rotor must be spun up to speed from rest (engagement) and slowed to a halt (disengagement). At the low-speed ends of these sequences the centrifugal moment is of a small magnitude but the aerodynamic moment can be enhanced by the adverse wind conditions and the blade can experience excessive flapping angles. This is known as blade sailing or, because of its potential to inflict damage to the upper tail boom, tunnel strike.

4.4.2 Lagging Motion

As already described, the helicopter rotor must attain a trimmed condition in forward flight. The disparity between the advancing and retreating sides of the disc is handled by the inclusion of flapping hinges on the rotor hub. Flapping motion introduces a phenomenon associated with a rotating system – that is, the rotor hub. This is the Coriolis acceleration.

4.4.3 Coriolis Acceleration

With reference to Figure 4.21, we have a circular disc on which is a radial direction rotating with it. The rotation will impart a velocity in a direction perpendicular to the radial direction (*the tangential direction*).

This velocity increases as the distance from the disc centre increases, so any point moving along the radial line (and remaining on that line) must acquire this increasing tangential velocity as it moves outwards from the disc centre. Obviously, if the movement is radially inwards towards the disc centre, this tangential velocity must decrease. We therefore have the situation where a rotating system causes a radial movement to acquire a velocity in a tangential direction which changes in magnitude – that is, an acceleration. This is the principle of the Coriolis acceleration.

Now, such acceleration will require the application of an appropriate force in the tangential direction. In many circumstances, such a force is *not* present and so the motion *cannot remain* on the line. In fact it will drift off in the opposite direction.

Coriolis can be found in many situations, one of which is the rotating winds found with depressions across the world.

Figure 4.22 shows the Earth with a region of low pressure whereby the air is moving towards it. The low pressure lies in the northern hemisphere and the inward movement of air in the direction towards the North Pole will be moving in towards the axis of the Earth's rotation. This will be subject to the Coriolis effect and since there is no force, the air will move in an easterly direction. Conversely, a wind moving away from the North Pole will be subject to a movement

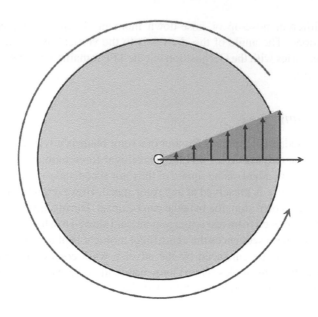

Figure 4.21 Rotating system

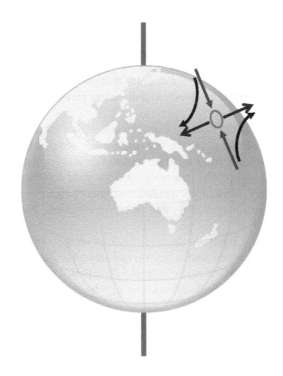

Figure 4.22 The effect of Coriolis on weather systems

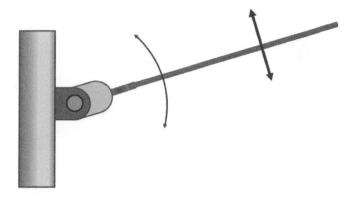

Figure 4.23 Blade flapping hinge

in a westerly direction. In this way a rotating flow will be generated. This gives the characteristic anticlockwise rotation around a low-pressure region in the northern hemisphere with the opposite rotation direction found in the southern hemisphere.

Figure 4.23 shows the typical flapping motion of a rotor blade. The rotation of the blade about the flapping hinge gives any point of the blade a motion which is radially inwards and outwards, depending on the blade position and its flapping motion. In fact an analysis of this situation shows that the Coriolis effect is proportional to the product of the flapping angle and the flapping rate, that is:

$$\text{Coriolis} \propto \beta\dot{\beta} \tag{4.10}$$

The normal condition for a rotor is with the rotor in a coned position upon which flapping oscillations are superimposed. This means that the Coriolis effect will be more pronounced for a rotor with a high coning angle.

It can now be seen that the flapping motion of a rotor blade generates a distribution of Coriolis forces on the blade, along its length, which combine to give an overall moment about an axis parallel to the rotor shaft (see Figure 4.23).

This will place a vibratory moment on the rotor hub structure which must be avoided. This can be achieved by allowing the blade to rotate in the rotor plane itself by installing a suitable hinge mechanism. This blade motion is termed lagging (Figure 4.24) and the instantaneous

Figure 4.24 Blade lag motion

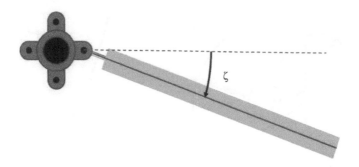

Figure 4.25   Definition of blade lag angle

blade position away from directly outwards is called the lag angle (see Figure 4.25) and is usually given the symbol ζ.

4.4.4   Lag Frequency

The blade lag motion is under the influence of centrifugal force – not unlike blade flapping motion; however, the moment arms of these two degrees of freedom are very different (see Figure 4.26).

Since the restoring moment arm is considerably smaller for lag motion compared with flap, it is the case that the natural lag frequency is smaller than the flapping frequency for the same hinge offset from the main rotor shaft. The two frequencies (normalized on the rotor speed) are:

$$\lambda_\beta = \sqrt{1 + \frac{3e}{2}}$$

$$\lambda_\zeta = \sqrt{\frac{3e}{2}}$$

(4.11)

where e is the hinge offset normalized on the rotor radius. The variation of these frequencies with hinge offset is shown in Figure 4.27.

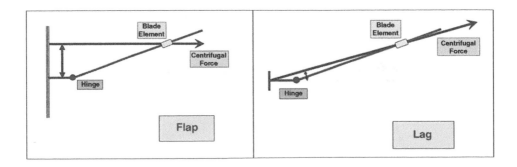

Figure 4.26   Restoring moment arm for flap and lag motion

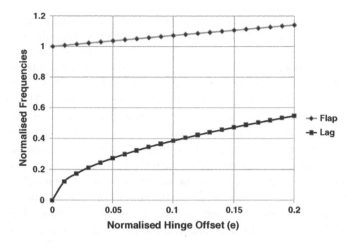

Figure 4.27   Variation of flap and lag frequencies

The first point to note is that the flapping frequency is always greater than unity with a zero offset giving exactly unity. This allows for a zero flapping hinge offset as seen with a teetering rotor. However, the lag frequency is below unity and equals zero for a zero hinge offset. This means that there must *always* be an offset for the lag hinge otherwise the rotor will not turn with the shaft. The fact that the lag frequency is less than unity opens the possibility of the helicopter suffering from a phenomenon called ground resonance. The reason for this is explained in section (4.4.6) of this chapter.

4.4.5   Blade Flexibility

The previous discussion has considered the rotor blades as rigid structurally but attached to the rotor hub by free hinges or flexures. In reality, the blades possess flexibility and can distort from the rigid straight shape adopted so far. The very significant tension under which the blades are subjected maintains an essentially straight shape but the complexities of the aerodynamic environment in which the blades operate will trigger elastic bending. How the blades do this is not appropriate for an introductory text. However, the blade natural modes are often used as a basis to define blade flexing and can be considered as separate for flap, lag and torsion (uncoupled modes) or when all three motions occur together (coupled modes). The torsion is of major importance for the KMax and Seasprite helicopters where this bending of the blades is used in rotor control – as previously discussed. The modal behaviour of the blades is closely involved in their design and, in the past, coupling of modes has been avoided. Modern blade development is now using blade modes, with their flap, lag and torsion components to enhance the aerodynamic performance. This use of favourable blade structural characteristics is known as 'aeroelastic tailoring'.

4.4.6   Ground Resonance

The provision of individual blade lag hinges means that the rotor head will now become more complicated, will require extra maintenance and will add to the weight and drag of the rotor head assembly. These are readily deduced; however, there is a dynamic implication of the blades moving in a lagwise sense. Firstly that the Coriolis effect connects the blade flapping and lagging motions. This can result in the two types of blade motion coupling together with implications for vibration and mechanical integrity. In addition, it is possible for the motion of the blades in lag to differ from blade to blade. If this should happen then the rotor, including blades, will have a CG which does not lie at the centre of the rotor. This will cause the generation of vibratory forces but also can cause a significant difficulty if the aircraft is sitting on the ground. The undercarriage will have an axial stiffness since it must compress on landing and therefore act as a spring. The stiffness of the undercarriage will allow the airframe itself to be able to vibrate at a frequency determined by the undercarriage stiffness and the appropriate moment of inertia of the airframe itself. We now have the situation of an oscillatory force being generated by the uneven blade lag motion, across the number of blades, acting on a system (the aircraft sitting on its undercarriage) which has its own natural frequencies. This is the classic combination which can result in a resonant condition if the forcing frequency closely approaches a natural frequency. In the case of a helicopter this can result in *ground resonance* which is a potentially devastating phenomenon and for which all new aircraft types are assessed before flight clearance is given.

Figure 4.28 shows the basic features governing the natural frequency of the aircraft on the ground and the forcing derived from uneven blade lag motion. Should the forcing frequency at the rotor head be close to the natural frequency of the fuselage rocking on its undercarriage, a resonant condition can occur.

The undercarriage of a helicopter can be of many forms and Figure 4.28 shows an example. The strut consists of a spring and damper, and this is directly connected to the wheel and tyre unit. The latter has spring and damping properties due to the tyre inflation. The spring contributes to the fuselage frequency while the damping is present to suppress any potential ground resonance. The suppression of ground resonance is crucial and will be discussed later in this chapter.

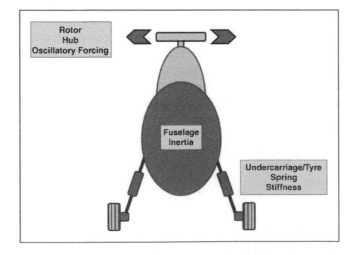

Figure 4.28   Factors for ground resonance

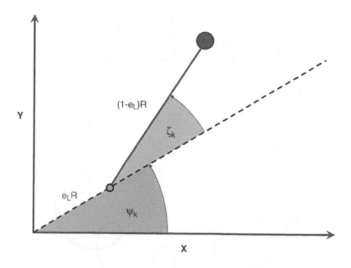

Figure 4.29   Simple model of ground resonance

4.4.6.1   Simple Analysis of the Problem – Rotor Only

The analysis of ground resonance can encompass a wide degree of complexity. The simplest model is to concentrate on the rotor alone and focus on the position of the CG. A further simplification is made where each blade is modelled by a concentrated mass joined to the lag hinge by a weightless rod. This is effectively viewing the individual blade mass centres.

The rotor/blade layout is shown in Figure 4.29.

The figure shows a single blade (of index k) where the rotor head (and the appropriate lag hinge) itself is at an azimuth angle of ψ_k. The blade itself is placed in a leading (forward lag) position with angle ζ_k.

The blade lag motion is then defined by:

$$\zeta_k = \zeta_0 \cos \lambda_\zeta \psi_k \tag{4.12}$$

which represents simple harmonic motion (SHM) of maximum amplitude ζ_0 and circular frequency λ_ζ, relative to the rotor speed.

Substituting this lag behaviour into the position of the blade masses, the centre of mass of the N blades can be calculated. The analysis is relatively straightforward and produces a result which, if interpreted as a single mass, shows a motion which can be described as tracing the petals of a flower – see Figure 4.30.

While this explains the CG motion, it is not helpful, so rather than the motion of one specific mass, the result can be interpreted as the motion of two equal masses. The values of radial location and circular frequency is shown in the following table:

	Mass 1	Mass 2
Mass	mN	mN
Radial location	$\dfrac{r_g \zeta_0}{2N} \cdot S_{+1}$	$\dfrac{r_g \zeta_0}{2N} \cdot S_{-1}$
Circular frequency	$(\lambda_\zeta + 1)\Omega$	$(\lambda_\zeta - 1)\Omega$

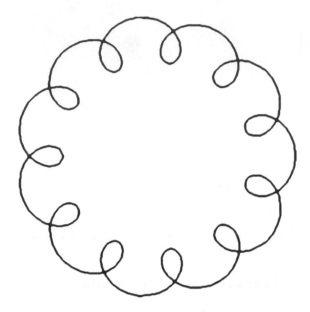

Figure 4.30   Motion path of centre of mass

where the S terms are as follows:

$$S_{\pm 1} = \frac{\sin[(\lambda_\zeta \pm 1)\pi]}{\sin[(\lambda_\zeta \pm 1)\pi/N]} \tag{4.13}$$

The variation of the S terms with non-dimensional lag frequency is shown in Figure 4.31. The frequency which is greater than the rotor speed is known as progressive, indicating that the mass is rotating, relative to the rotor, in the direction of rotor rotation. The other is known as regressive since it moves against the rotor direction.

Figure 4.32 shows the result for the case of a normalized lag hinge offset of 0.1. The rotor has four blades and the two masses are circles at 2 and 8 o'clock. The overall rotor CG is the circle in between and, since the two masses are equal, it lies at their mid-point.

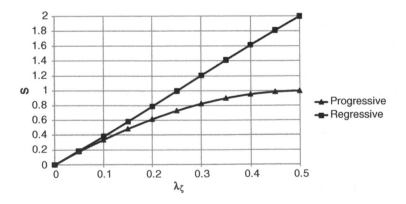

Figure 4.31   Radial location of rotating masses

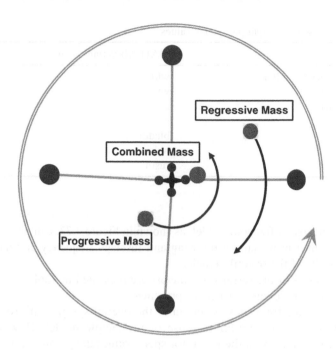

Figure 4.32 Twin masses of rotor

4.4.6.2 Simple Analysis with Fuselage Motion

The analysis of a simple model – see Figure 4.33 – can only measure the natural frequencies of the rotor. This is perfectly adequate if the rotor speed values capable of inducing a resonance are required. It cannot give any indication of the damping necessary to suppress any instability. A full 5/6 degree of freedom (DOF) model must be used for the most exacting calculations;

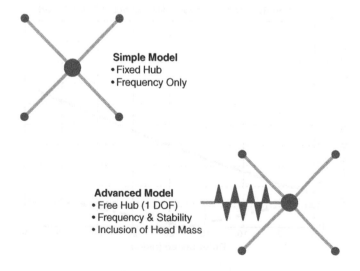

Figure 4.33 Ground resonance models

Table 4.1 Ground resonance input parameter values

Parameter	MATLAB variable name	Value
Static rotor lag frequency (rad/s)	wl0	15.22
Fuselage frequency (rad/s)	wy	12
Lag hinge offset (m)	e	1.22
Rotor radius (m)	r	6.4
Blade mass (kg)	bladm	24.80
Number of blades	nblad	4
Fuselage mass (kg)	fusm	500

however, a single degree of freedom model, as shown in Figure 4.33, can provide very useful insight into the manner in which the rotor lag and fuselage frequency and damping values influence the stability of the overall vehicle.

The numerical values of the various parameters are presented in Table 4.1.

From these data the following results are presented.

Figure 4.34 shows the frequency variation of the overall system with rotor speed. The shape of the graphs is very similar to those of the simple model. However, there are important differences. Firstly, at the low rotor speed values the graphs show an interesting behaviour in that they seem to exchange characteristics. This is a common situation when two uncoupled systems (in this case the blade lagging and the fuselage motion) are connected. The second, and more important, observation is the two regressing modes in the region of 35 rad/s rotor speed. They coalesce, which cannot happen in a simple analysis of frequency only. Consulting Figure 4.35 shows that an undamped system displays instability (positive real part) for this rotor speed range. It is neutrally damped at other rotor speeds.

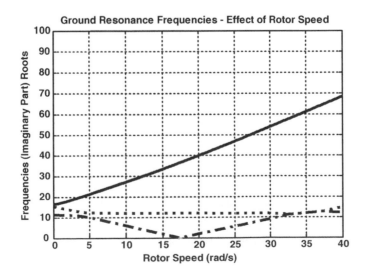

Figure 4.34 Basic frequency variation (imaginary part) with rotor speed

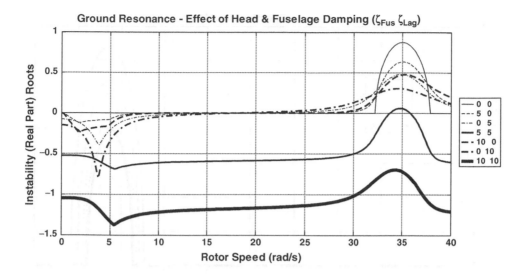

Figure 4.35　Stability variation (real part) with rotor speed

4.4.6.3　Effect of Damping

There is therefore a potential instability which must be suppressed. This usually is accomplished by adding damping and, while it is possible to incorporate this damping in several parts of the airframe, in order to fully suppress ground resonance, certain requirements must be met.

To see the effect, a range of damping is applied to the rotor head and the fuselage via the undercarriage.

Figure 4.35 also shows the effect of adding damping to the rotor lag (legend right-hand column) and the fuselage (legend left-hand column). If damping is added to either the rotor or the fuselage while leaving the other undamped, instability remains. If the damping is applied to both rotor and fuselage, then elimination of the instability can be achieved. This result is predicted in Coleman and Feingold [2] where they express this situation thus. If ζ_{FUS} is the damping ratio for the fuselage and ζ_{ROT} is that for the rotor then the following condition must apply for stability:

$$\zeta_{FUS} \cdot \zeta_{ROT} \geq h > 0 \qquad (4.14)$$

The product of the two damping values must be positive, which cannot be achieved if one is undamped. Therefore damping must be applied evenly, which is not always convenient.

With reference to Figure 4.36, the influence of the rotor head mass can be seen. For the simple model, the rotor head was fixed and, therefore, the fuselage could take no part in the motion and thus was isolated from the rotor. Now the rotor can move laterally, the fuselage is included in the motion, and the effect can be expressed via the effective head mass. The analysis of the system was carried out using an energy-based method. Therefore the fuselage must be included via its kinetic energy. This calls into play the manner in which the fuselage moves. If the fuselage motion is considered to be in roll only, then the motion about the CG will be a combination of sway and roll angle. It will have, therefore, a point about which it is effectively rotating. This point will govern the interaction between the fuselage dynamics and the rotor blade lag motion. Figure 4.36 shows that the lower the value of the effective head mass, the more unstable the ground resonance motion. Figure 4.37 shows two situations where the rotation centre is placed low or high on the fuselage. For a given rotation amount, the low centre gives more head

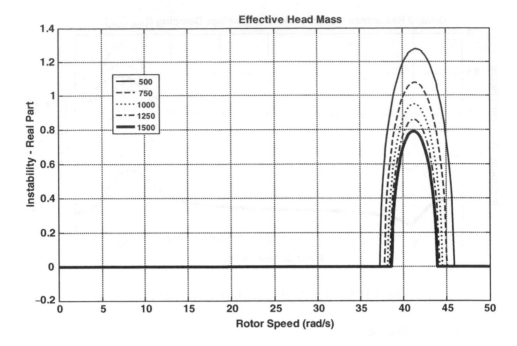

Figure 4.36   Stability variation – influence of effective head mass

movement. If the head mass is to reflect the kinetic energy of the complete aircraft then the situation with greater head motion will give rise to a lower value of effective head mass. Also the greater head motion will encourage the interaction between the blade lagging and the motion of the fuselage on its undercarriage. Hence the lower head mass can be expected to give a greater amount of feedback and thus a more unstable condition can be predicted.

Figure 4.38 shows the effect of fuselage frequency giving more instability for the higher values.

Figure 4.39 shows the effect of blade lag frequency and the higher values of lag stiffness giving the more stable condition. Therefore, a semi-rigid rotor system will require smaller amounts of damping to ensure stability.

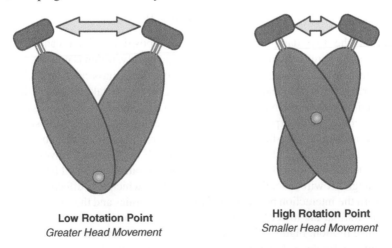

Figure 4.37   Location of fuselage rotation

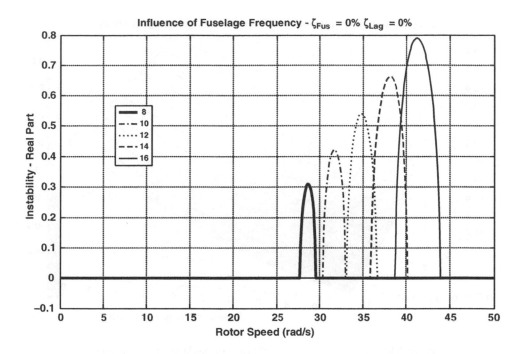

Figure 4.38   Stability variation – influence of fuselage frequency

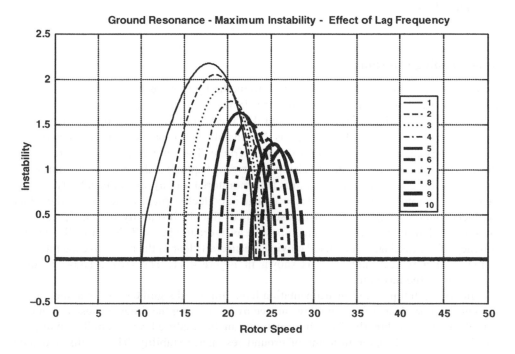

Figure 4.39   Stability variation – influence of lag frequency

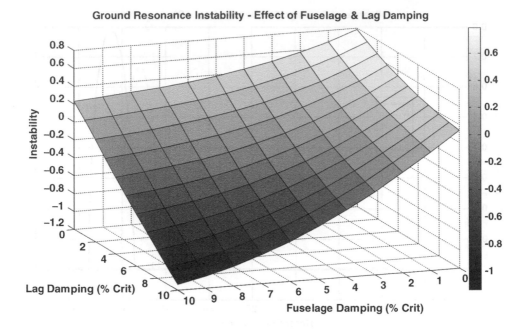

Figure 4.40   Stability variation – influence of fuselage and lag damping

Figure 4.40 is a surface plot showing the effect on the stability of the aircraft for a range of lag and fuselage damping values. The shape of the surface is another illustration of the law defined in Equation 4.14.

4.4.6.4   Testing Procedures

When a new helicopter design has been built to the first prototype stage it is usual to verify that it is not susceptible to ground resonance. In the past, the airframe was stripped to a minimum all-up weight. This uses the result that the lighter the helicopter, the greater the ground resonance vibration. The aircraft is then installed in a rig – known as a snatch rig – which attaches cables to strong points on the airframe. The rotor is gradually increased in speed and the pilot applies a cyclic pitch input to trigger flapping and hence lagging motion of the blades. The behaviour of the aircraft is then monitored for any sign of instability. This is repeated over a rotor speed range and also a range of collective pitch values so that the effect of reduced undercarriage load can be ascertained. Should instability be anticipated, the cables are pulled tight by a set of actuators such as hydraulic jacks. This causes the natural frequency of the fuselage to be raised moving the aircraft clear of the regressive lag mode. A schematic of a typical snatch rig is shown in Figure 4.41. The arrows show possible attachments to the airframe.

Prediction techniques and instrumentation have improved significantly since then and a helicopter can be cleared for ground resonance by monitoring the timewise fuselage and/or rotor blade motion directly. This information can be analysed very rapidly giving the condition of the helicopter in terms of ground resonance stability. The results will then

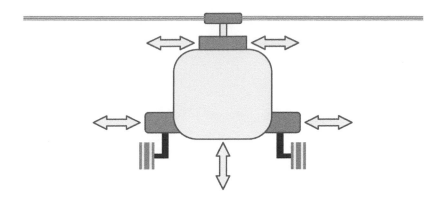

Figure 4.41   Snatch rig

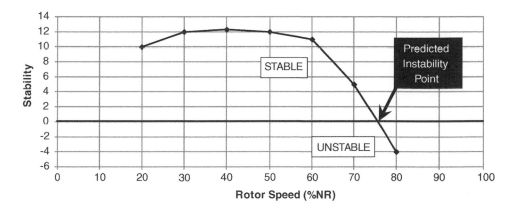

Figure 4.42   Example plot of locating incipient ground resonance

give the stability variation with respect to rotor speed. Figure 4.42 shows a possible variation and the behaviour of the curve approaching the axis gives an indication of an incipient instability.

References

1. Saunders, G.H. (1975) *Dynamics of Helicopter Flight*, John Wiley & Sons, Inc.
2. Coleman, R.P. and Feingold, A.M. (1958) Theory of self-excited mechanical oscillations of helicopter rotors with hinged blades, NACA Report TN-1351.

5

Rotor Aerodynamics in Forward Flight

The aerodynamic situation in forward flight is complex. Numerical methods have largely taken over the task of evaluation but an analytical treatment, using simplifying assumptions, is valuable for providing a basic understanding of rotor behaviour. Such a treatment is the subject of this present chapter. The mechanisms of the previous chapter affect essentially the details of blade element theory. Before turning to that, however, it is useful to examine briefly what can be made of momentum theory, which as has been said is principally a theory for hover and axial flight; also it may be asked to what situations one is led in considering a more detailed wake analysis under forward flight conditions.

5.1 Momentum Theory

As the rotor begins to attain forward flight speed, the velocity of the air entering the rotor will increase as the forward flight speed combines with the sum of the induced velocity and any axial flight speed. In addition, it will approach the rotor disc at an inclined angle, hence any development of the actuator disc theory to include forward flight must address these issues. The type of streamtube shape seen in forward flight with an actuator disc is shown in Figure 5.1.

The modelling of the actuator disc theory into forward flight was addressed by Glauert [1] who devised a scheme which was based on the following:

- The concept of the streamtube is adopted.
- The area presented to the airflow for mass flow calculations is the whole disc area. This ensures the method is consistent with lifting line theory for a rectangular wing. *In this case, the area for momentum calculations is a circle with the wing as diameter.*
- The induced velocity is normal to the disc plane.
- The induced velocity in the far wake is twice that at the rotor disc – as in axial flight – and is also normal to the disc plane.

Basic Helicopter Aerodynamics, Third Edition. John Seddon and Simon Newman.
© 2011 John Wiley & Sons, Ltd. Published 2011 by John Wiley & Sons, Ltd.

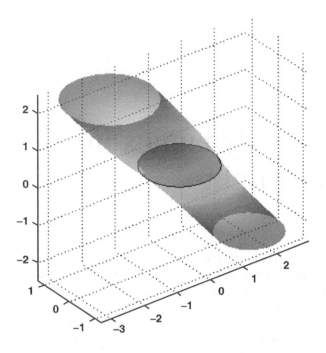

Figure 5.1   Streamtube in forward flight

The velocities of the air through the streamtube are shown in Figure 5.2.
  Expressing all velocities relative to the rotor:

- Far upstream, the airflow is horizontal.
- At the rotor disc, the induced velocity is added (*vectorially*) to the forward flight velocity.
- Far downstream, twice the induced velocity is added (*vectorially*) to the forward flight velocity.

With the notation of Figure 5.2, and noting the second statement of the scheme, the mass flow into the rotor disc is given by:

$$\rho A \sqrt{V_X^2 + (V_Z + V_i)^2} \qquad (5.1)$$

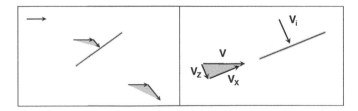

Figure 5.2   Flow directions and velocity components for momentum theory in forward flight

This gives the rate of change of momentum normal to the rotor disc as:

$$\rho A \sqrt{V_X^2 + (V_Z + V_i)^2} \cdot 2V_i \qquad (5.2)$$

This must equal the rotor thrust, so we find the following expression:

$$V_i = \frac{T}{2\rho A} \cdot \frac{1}{\sqrt{V_X^2 + (V_Z + V_i)^2}} \qquad (5.3)$$

This equation can be normalized so that advance ratio components are used:

$$\lambda_i = \frac{T}{2\rho A} \cdot \frac{1}{V_T^2} \cdot \frac{1}{\sqrt{\mu_X^2 + (\mu_Z + \lambda_i)^2}}$$

$$= \frac{C_T}{4} \cdot \frac{1}{\sqrt{\mu_X^2 + (\mu_Z + \lambda_i)^2}} \qquad (5.4)$$

The thrust coefficient dependence can be incorporated by normalizing the advance ratio components and the induced velocity by the hover induced velocity λ_{i0}:

$$\lambda_{i0} = \sqrt{\frac{T}{2\rho A}} \qquad (5.5)$$

that is:

$$\bar{\mu}_X = \frac{\mu_X}{\lambda_{i0}}$$

$$\bar{\mu}_Z = \frac{\mu_Z}{\lambda_{i0}} \qquad (5.6)$$

$$\bar{\lambda}_i = \frac{\lambda_i}{\lambda_{i0}}$$

Consequently, Equation 5.4 becomes:

$$\bar{\lambda}_i = \frac{1}{\sqrt{\bar{\mu}_X^2 + (\bar{\mu}_Z + \bar{\lambda}_i)^2}} \qquad (5.7)$$

which can be recast as:

$$\bar{\mu}_X^2 + (\bar{\mu}_Z + \bar{\lambda}_i)^2 = \left(\frac{1}{\bar{\lambda}_i}\right)^2 \qquad (5.8)$$

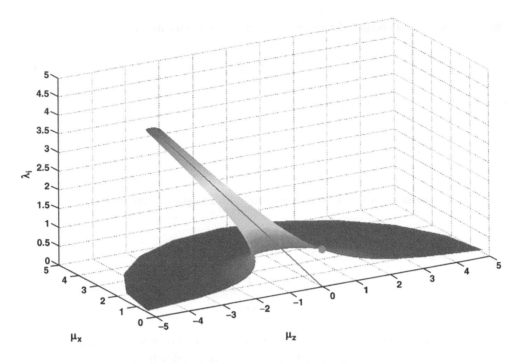

Figure 5.3　Solution surface for actuator disc theory

Hence, for a given value of downwash, that is in a given plane parallel to the base (μ_X, μ_Z) plane, Equation 5.8 defines a circle:

$$\text{Centre} \Rightarrow (-\overline{\mu}_Z, 0)$$

$$\text{Radius} \Rightarrow \frac{1}{\overline{\lambda}_i} \tag{5.9}$$

The surface defined by (5.8) is shown in Figure 5.3.

The figure shows both the basic shape of axial flight and the effect of forward flight. As the forward flight velocity increases, the problems of moderate descent rate diminish and eventually disappear. The figure shows the mathematical results of the theory, but in reality the effect of forward speed is of providing a second mechanism for the rotor to clear the vorticity forming at the rotor disc.

The usual situation is of wanting to determine the value of the downwash from the advance ratio components and the rotor thrust. This means solving Equation 5.7, if we assume that we are working with normalized values of velocity.

This can be solved in two ways. Firstly it can be viewed as an iterative scheme, where an initial guess at a value of $\overline{\lambda}_i$ is substituted into the right-hand side of (5.7) which gives a revised value. This has been shown to work well for high speeds where the $\overline{\mu}_X$ term dominates. However, at low speeds, the calculation can take a large number of iterations to converge on a value to a sensible degree of accuracy. In hover, the routine will never converge – unless the correct solution is used at the outset. A modification to the method was devised by Hansford who re-expressed (5.7) as:

$$\bar{\lambda}_i - \frac{1}{\sqrt{\bar{\mu}_X^2 + (\bar{\mu}_Z + \bar{\lambda}_i)^2}} = 0 \tag{5.10}$$

This is now the solution for the root of a function of $(\bar{\lambda}_i)$ that is equal to zero – and invoking Newton–Raphson. This gives the following iterative scheme:

$$\bar{\lambda}_i \Leftarrow \bar{\lambda}_i - \left\{ \frac{\bar{\lambda}_i - \left[1 \Big/ \sqrt{\bar{\mu}_X^2 + (\bar{\mu}_Z + \bar{\lambda}_i)^2} \right]}{1 + \left\{ (\bar{\mu}_Z + \bar{\lambda}_i) \Big/ \left[\bar{\mu}_X^2 + (\bar{\mu}_Z + \bar{\lambda}_i)^2 \right]^{3/2} \right\}} \right\} \tag{5.11}$$

The calculation is more complicated but convergence is achieved efficiently over the entire speed range.

Equation 5.10 can be squared and rearranged to give:

$$\bar{\lambda}_i^4 + 2\bar{\mu}_Z \cdot \bar{\lambda}_i^3 + \left(\bar{\mu}_X^2 + \bar{\mu}_Z^2 \right) \cdot \bar{\lambda}_i^2 - 1 = 0 \tag{5.12}$$

This is a quartic which gives four solutions. These comprise a complex conjugate pair, which can be discarded. The two real solutions are of opposite sign. As the induced velocity is always taken to be positive, the negative value can also be discarded. Hence the real positive solution is the one that should be used. This calculation has been performed over a range of values for $\bar{\mu}_X$ and $\bar{\mu}_Z$ and the results are shown in Figure 5.4.

5.2   Descending Forward Flight

As can be seen in Figure 5.4, the majority of the surface is the same as Figure 5.3; however, the main difference is in the vortex ring state region of moderate descent with low to zero forward flight speed. Here the solution experiences a rapid jump in value – which, in fact, is not totally incorrect. Perhaps the most significant error in character is the fact that a steady value of induced velocity is indicated – the theory defines this – but, as already discussed, this is not the case in reality. We therefore need to define a region dividing those flight conditions which can be handled sensibly by actuator disc theory (ADT) and those which require a different approach. This requires the introduction of a vortex ring state boundary. This will be governed by the ability of the rotor to clear the vorticity which is being generated at the rotor disc.

So if we consider the overall advance ratio:

$$\bar{\mu} = \sqrt{\bar{\mu}_X^2 + (\bar{\mu}_Z + \bar{\lambda}_i)^2} \tag{5.13}$$

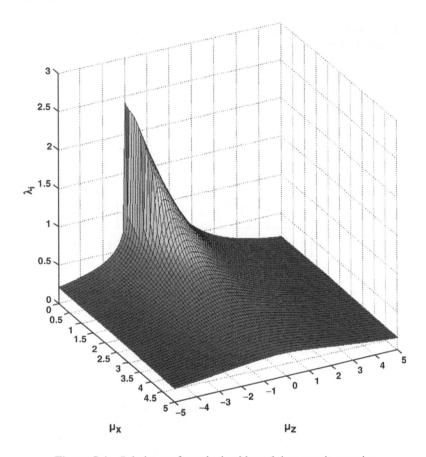

Figure 5.4   Solution surface obtained by solving quartic equation

then that is the total normalized velocity through the rotor disc and will be the main influence in carrying the vorticity away. Work conducted on this topic by Newman *et al.* [2] defined this velocity to have a critical value, whence the boundary perimeter is defined by:

$$\bar{\mu}_{\text{CRIT}} = \sqrt{\bar{\mu}_X{}^2 + \left(\bar{\mu}_Z + \bar{\lambda}_i\right)^2} \tag{5.14}$$

This formula assumes that the forward and vertical velocities are completely efficient in disposing of this vorticity. However, consultation of the flow visualization of Drees [3] by Perry [4], showed that the forward velocity component does not operate with perfect efficiency. In addition, Brand [5] proposed a similar adjustment to the total vertical component. The revised boundary definition now becomes:

$$\bar{\mu}_{\text{CRIT}} = \sqrt{k_1^2 \cdot \bar{\mu}_X{}^2 + k_2^2 \cdot \left(\bar{\mu}_Z + \bar{\lambda}_i\right)^2} \tag{5.15}$$

which can be rearranged to give:

$$\bar{\mu}_Z + \bar{\lambda}_i = \pm \frac{\sqrt{\bar{\mu}_{CRIT}{}^2 - k_1^2 \cdot \bar{\mu}_X{}^2}}{k_2}$$

$$\bar{\mu}_Z = \pm \frac{\sqrt{\bar{\mu}_{CRIT}{}^2 - k_1^2 \cdot \bar{\mu}_X{}^2}}{k_2} - \bar{\lambda}_i$$

(5.16)

If we now substitute (5.16) in (5.7) we obtain:

$$\bar{\lambda}_i = \frac{1}{\sqrt{\bar{\mu}_X{}^2 + \dfrac{1}{k_2^2}\left(\bar{\mu}_{CRIT}{}^2 - k_1^2 \cdot \bar{\mu}_X{}^2\right)}}$$

(5.17)

$$= \frac{k_2}{\sqrt{\bar{\mu}_X{}^2\left(k_2^2 - k_1^2\right) + \bar{\mu}_{CRIT}{}^2}}$$

Hence, for a given value of $\bar{\mu}_{CRIT}$ and $\bar{\mu}_X$ the values of $\bar{\mu}_Z$ and $\bar{\lambda}_i$, the boundary point, can be defined using (5.16) and (5.17).

Using the following values:

$$\bar{\mu}_{CRIT} = 0.74$$

$$k_1 = 0.65$$

$$k_2 = 0.9$$

(5.18)

the boundary is as shown in Figure 5.5.

As can be seen, the boundary hangs – like a necklace – around the area of concern. Having established the boundary, outside of which ADT can be sensibly used, the situation within the boundary needs comment. This is the region where the flow is dominated by vorticity and where a much more detailed theory needs to be used. Alternatively, which was the way initially used, experimental data can be introduced. Both experiment and more detailed analysis (see e.g. Bramwell's Chapter 4) confirm that the Glauert proposal works well.

In practice the induced velocity cannot be expected to be constant over the area of the disc. Standard aerofoil theory would suggest an upwash at the leading edge and a greater-than-mean downwash at the trailing edge. To allow for a variation of this kind, Glauert proposed a second formula:

$$V_i(x, \psi) = V_{i0}(1 + Ex \cos \psi)$$

(5.19)

where V_{i0} is the value at the centre, taken to be that given by (5.3), x is the non-dimensional radius from the centre and ψ is the azimuth angle. If the constant E is chosen to be greater than 1.0 (typically 1.2), the formula gives a negative value, that is an upwash, at the leading edge ($\psi = 180°$). Equation 5.19 is often used as an input to numerical methods.

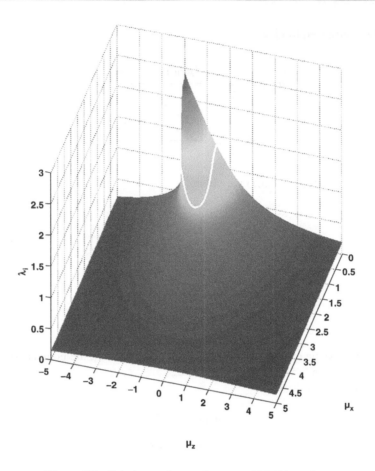

Figure 5.5   Solution surface and proposed VRS boundary

A plot of this profile, with E having the value 1.2, is shown in Figure 5.6.

The plot is an inclined plane; note the negative values at the front of the disc. A justification for this downwash model can be found in examining a more detailed model using vortex rings, if they are positioned in an inclined stack, as shown in Figure 5.7.

The resulting downwash variation is shown in Figure 5.8. As can be seen, the planar nature is shown within the central part of the disc and a negative value is predicted at the front of the disc. The ridge of the distribution in a region close to the disc edge is a local effect attributed to the vortex ring closest to the rotor disc.

More elaborate treatments of the non-uniform induced velocity in forward flight have been devised, among which one of the foremost is the method of Mangler and Squire [6]. Described at length by Bramwell (p. 127ff.), this method has shown satisfactory agreement with controlled experiments and is stated to be very useful in rotor calculations.

Reverting to the Glauert formula for uniform induced velocity, Equation 5.3, the induced power is:

$$P_i = TV_i = \frac{T^2}{2\rho A\sqrt{V_X^2+(V_Z+V_i)^2}} \tag{5.20}$$

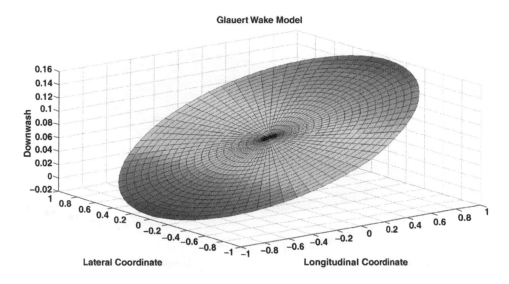

Figure 5.6 Glauert downwash model

which at normal forward flight speeds becomes approximately:

$$P_i = TV_i = \frac{T^2}{2\rho A V}$$

$$= T\left(\frac{T}{A}\right)\frac{1}{2\rho V} \qquad (5.21)$$

$$= \frac{Tw}{2\rho V}$$

that is, directly proportional to the disc loading w.

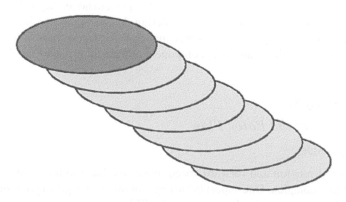

Figure 5.7 Stack of pennies visualization of vortex ring geometry

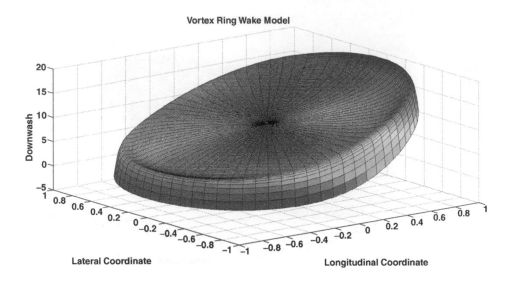

Figure 5.8   Vortex ring geometry, downwash variation

In non-dimensional terms the first equality of (5.20) is simply:

$$C_{Pi} = \lambda_i \cdot C_T \tag{5.22}$$

where λ_i is $V_i/\Omega R$. It will be useful for the forward flight case to adopt a suffix 'i' for that part of the total induced flow which is due to the thrust-dependent induced velocity V_i, as distinct from a part due to the forward velocity V.

As with hover, a practical approximation to allow for the effect of non-uniformity in V_i and other smaller correction factors is obtained by applying an empirical factor k such that:

$$C_{Pi} = k_i \cdot \lambda_i \cdot C_T \tag{5.23}$$

The value of k_i in forward flight is somewhat higher than that in hover, say 1.20 compared with the formerly suggested 1.15 (Section 3.6). Countering this, however, the induced velocity is seen in Figure 5.5 to become quite small even at moderate forward speeds: it will duly emerge that C_{Pi} is then much smaller than other components of the total power requirement.

5.3   Wake Analysis

5.3.1   Geometry of the Rotor Flow

5.3.1.1   Sweep Angle

The combination of rotation and forward speed makes the local inflow distribution across a helicopter rotor disc complex. One aspect is the angle of inflow at a given point of a rotor blade at a given azimuth angle; that is, the sweep angle. This section derives, in closed form, the equation defining the sweep angle contours.

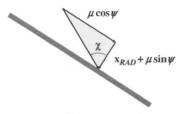

Figure 5.9   Local velocity components

Derivation of Contours

The rotor dimension is normalized to a unit radius. Also, the velocity components are normalized on the rotor tip speed. A general point of a rotor blade is at a normalized radius of (x_{RAD}) and azimuth angle (ψ).

The velocity components defining the local sweep angle are shown in Figure 5.9.

From the figure, the sweep angle is defined by:

$$\tan \chi = \frac{\mu \cos \psi}{x_{\text{RAD}} + \mu \sin \psi} \tag{5.24}$$

In order to determine the contour equations we use the following axes system (see Figure 5.10): *the X axis (abscissa) lies in the incident airflow due to forward speed, that is over the tail, and the y axis (ordinate) lies to starboard.* With this axis system, the relationship between the normalized rotor radius (x_{RAD}) and the azimuth angle (ψ) is as shown in Figure 5.11.

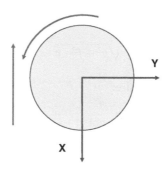

Figure 5.10   Axes system definition

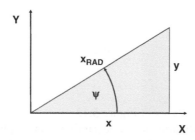

Figure 5.11   Transformation between rotor polar and Cartesian coordinates

The transformation equations are:

$$
\begin{aligned}
x_{RAD} &= \sqrt{x^2 + y^2} \\
\cos\psi &= x/x_{RAD} \\
\sin\psi &= y/x_{RAD}
\end{aligned}
\tag{5.25}
$$

Substituting (5.25) into (5.24) and clearing fractions gives:

$$
x_{RAD} + \mu \cdot \frac{y}{x_{RAD}} = \mu \cdot \cot\chi \cdot \frac{x}{x_{RAD}}
\tag{5.26}
$$

hence:

$$
x^2 + y^2 - \mu \cdot \cot\chi \cdot x + \mu y = 0
\tag{5.27}
$$

The equation of a general circle is given by:

$$
x^2 + y^2 + 2gx + 2fy + c = 0
\tag{5.28}
$$

where the centre of the circle is:

$$
(-g, -f)
\tag{5.29}
$$

and the radius is:

$$
\sqrt{g^2 + f^2 - c}
\tag{5.30}
$$

From (5.27)–(5.29), the circle has centre:

$$
\left(\frac{\mu}{2} \cdot \cot\chi, -\frac{\mu}{2} \right)
\tag{5.31}
$$

and radius:

$$
\sqrt{\left(\frac{\mu}{2}\cot\chi \right)^2 + \left(-\frac{\mu}{2} \right)^2}
\tag{5.32}
$$

which reduces to:

$$
\frac{\mu}{2} \csc\chi
\tag{5.33}
$$

Geometrical Construction
This can be shown geometrically as in Figure 5.12.

The construction begins with drawing a line parallel to the flight direction (x) positioned at the distance of half the advance ratio; this is in fact passing through the centre of the reverse flow region. A second line is constructed from the origin (O, the centre of the rotor) at an angle to the

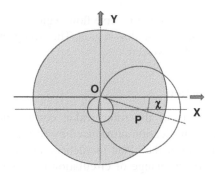

Figure 5.12 Geometrical construction of sweep angle contour

abscissa equal to the sweep angle contour required (χ). This intersects the first construction line at point P. Using P as the centre and OP the radius, a circle is drawn. The part of this circle within the rotor disc (the unit circle) is the required contour. An example of these contours is shown in Figure 5.13 for an advance ratio of 0.25.

Examination of the equation governing the sweep angle contours enables the following conclusions to be made:

- The sweep angle contour is a circle.
- The zero angle contour is the advancing blade ($\psi = 90°$) and the retreating blade ($\psi = 270°$).

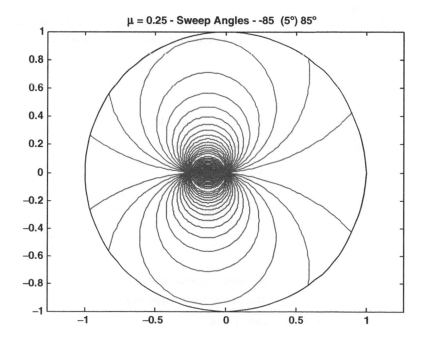

Figure 5.13 Sweep angle contours

- The 90° contour is the periphery of the reverse flow region.
- All the contours pass within the reverse flow region. The sweep angle remains the same, but the flow direction is completely reversed.

5.3.1.2 Blade Vortex Interaction

As concerns a detailed analysis of the rotor wake, corresponding to that outlined in Chapter 2 for the hover, the complication introduced by forward flight comes down to the fact that at a given radial position, the blade incidence, and hence the circulation, varies widely around the azimuth. Each change of circulation results in a counter vortex being shed into the wake and since the change is a circumferential one, the vortex line in this case lies in the spanwise direction. This system of 'shed' vortices is now additional to the 'trailing' vortex system arising, as in hover, from the spanwise variations in circulation.

Undeterred by such multiplicity of complication, the modern computer, guided by skilled workers among whom pioneers include Miller, Piziali and Landgrebe, is still capable of providing solutions and indeed building year upon year. The power of the modern computer is releasing rotor analysis from restrictions which were necessary in the past to achieve a realistic result. An example from Landgrebe's calculations shows in Figure 5.14 a theoretical wake boundary at low advance ratio, compared with experiment by smoke visualization and also with the Glauert momentum theory solution. The numerical solution and the experimental evidence agree well; momentum theory gives a much less accurate picture. A feature to note is that the boundary at the front of the disc lies close to the disc. This is illustrated in Figure 1.22c. At a higher advance ratio, more representative of forward flight, this feature and the general sweeping back of the wake would be much more marked.

This brief reference to what is a large subject in itself will suffice for the purposes of the present book. Extended descriptions can be found in the standard textbooks.

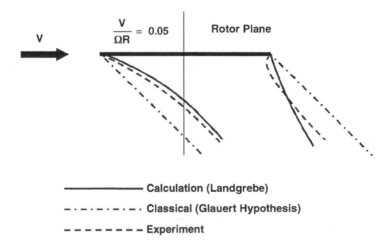

Figure 5.14 Wake boundaries at low advance ratio (after Landgrebe)

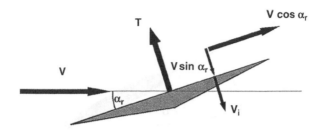

Figure 5.15 Disc incidence and component velocities in forward flight

5.4 Blade Element Theory

5.4.1 Factors Involved

An exposition of blade element theory follows the same broad lines as used for hover (Chapter 3), taking into account, however, the extra complexities involved in forward flight. We begin by introducing the additional factors which enter into a forward flight condition.

Figure 5.15 shows a side view of the rotor disc – strictly a shallow cone as we have already seen. Motion is to the left and is assumed horizontal; that is to say, without a climb component. The plane enclosing the edge of the disc – the tip-path plane (TPP) – makes an angle α_r with the oncoming stream direction. α_r is reckoned positively downwards since that is the natural direction of tilt needed to obtain a forward component of the thrust necessary to maintain forward flight. We shall use small-angle approximations as required. The flight velocity V has components $V \cos \alpha_r$ and $V \sin \alpha_r$ along and normal to the TPP. The advance ratio is given by:

$$\mu = \frac{V \cos \alpha_r}{\Omega R} \simeq \frac{V}{\Omega R} \tag{5.34}$$

as used previously. The total inflow through the rotor is the sum of $V \sin \alpha_r$ and V_i, the thrust-related induced velocity.

Referring to Figure 5.16, the resultant velocity U at a blade section is now a function of rotor rotation (i.e. rotor azimuth), helicopter forward speed, induced velocity and blade flapping motion. Components of U in the plane of the rotor rotation are U_T and U_P; additionally, because of the forward speed factor there is a spanwise component U_R, shown in Figure 5.17. Components U_T and U_R are readily defined; to first order these are:

$$U_T = \Omega r + V \sin \psi \tag{5.35}$$

$$U_R = V \cos \psi \tag{5.36}$$

or, in non-dimensional form:

$$u_T = x + \mu \sin \psi \tag{5.37}$$

$$u_R = \mu \cos \psi \tag{5.38}$$

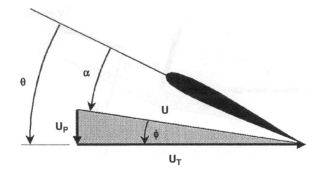

Figure 5.16   Component velocities U_T and U_P

The component U_P has three terms, non-dimensionally, as follows:

1. The inflow factor:

$$\lambda = \frac{V \sin \alpha_r + V_i}{\Omega R} \simeq \mu \alpha_r + \lambda_i \qquad (5.39)$$

2. A component of u_R normal to the blade, which for a flapping angle β relative to the reference plane is seen (Figure 5.18) to be βu_R or:

$$\beta \cdot \mu \cos \psi \qquad (5.40)$$

3. A component resulting from the angular motion about the flapping hinge; at station r along the span, this is:

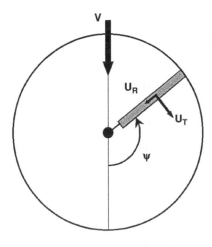

Figure 5.17   Component velocities U_T and U_R

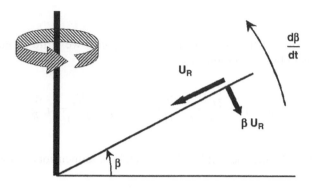

Figure 5.18 Flapping term in U_P

$$r \cdot \frac{d\beta}{dt} = r \cdot \frac{d\beta}{d\psi} \cdot \frac{d\psi}{dt}$$

$$= r \cdot \Omega \cdot \frac{d\beta}{d\psi}$$

(5.41)

when non-dimensionalising with respect to tip speed this becomes:

$$\frac{1}{\Omega R}\left(r \cdot \Omega \cdot \frac{d\beta}{d\psi}\right) = \frac{r}{\Omega R} \cdot \Omega \cdot \frac{d\beta}{d\psi}$$

$$= \frac{r}{R} \cdot \frac{d\beta}{d\psi}$$

$$= x \cdot \frac{d\beta}{d\psi}$$

(5.42)

Thus, combining these terms together,

$$u_P = \lambda + \beta\mu\cos\psi + x \cdot \frac{d\beta}{d\psi}$$

(5.43)

For small angles the resultant velocity U may be approximated by U_T. The blade angle of incidence may be written as:

$$\alpha = \theta - \phi = \theta - \frac{U_P}{U_T} = \theta - \frac{u_P}{u_T}$$

(5.44)

Note that whereas the values of θ and ϕ depend upon the choice of reference plane, the actual blade incidence α does not, so the expression $(\theta - u_P/u_T)$ is independent of the reference plane used.

5.4.2   Thrust

Following the derivation of the hover analysis in Chapter 3 we write an elementary thrust coefficient of a single blade at station r as:

$$dC_{T\text{ BLADE}} = \frac{\frac{1}{2}\rho U^2 \cdot c\,dr \cdot C_L}{\frac{1}{2}\rho \cdot \pi R^2 \cdot (\Omega R)^2} = \frac{c}{\pi R} \cdot \frac{U_T^2}{(\Omega R)^2} C_L \frac{dr}{R} \tag{5.45}$$

and for N blades, introducing the solidity factor s and non-dimensionalising,

$$dC_T = s \cdot u_T^2 C_L\,dx \tag{5.46}$$

On expressing C_L in the linear form (i.e. the blade is unstalled):

$$C_L = a\alpha = a\left(\theta - \frac{u_P}{u_T}\right) \tag{5.47}$$

from which (5.46) becomes:

$$dC_T = sa\left(\theta u_T^2 - u_P u_T\right)dx \tag{5.48}$$

For the hover we were able to write $u_T = x$, $u_P = \lambda$: in forward flight, however, u_T and u_P, and in general θ also, are functions of azimuth angle ψ. The elementary thrust must therefore be averaged around the azimuth and integrated along the blade. It is convenient to perform the azimuth averaging first and we therefore write the thrust coefficient of the rotor as:

$$C_T = \int_0^1 sa\left[\frac{1}{2\pi}\int_0^{2\pi}\left(\theta u_T^2 - u_P u_T\right)d\psi\right]dx \tag{5.49}$$

To expand the terms within the inside brackets, we recall from Chapter 4 that the flapping angle β may be expressed in the form:

$$\beta = a_0 - a_1\cos\psi - b_1\sin\psi \tag{5.50}$$

from which also we have:

$$\frac{d\beta}{d\psi} = a_1\sin\psi - b_1\cos\psi \tag{5.51}$$

For the feathering angle θ a similar Fourier expansion (Equation 4.9) can be used: however, there is always one plane, the plane of the swashplate or no-feathering plane (NFP), relative to which there is no cyclic change in θ; for our analytical solution therefore this will be used as the reference plane. Thus we have $\theta = \theta_0$, constant in azimuth, and following the same procedure as for hover we shall assume an untwisted blade, giving θ_0 constant also along the span. Averaging round the azimuth will make use of the following results:

$$\int_0^{2\pi} \sin\psi \, d\psi = 0$$

$$\int_0^{2\pi} \cos\psi \, d\psi = 0$$

$$\int_0^{2\pi} \sin\psi\cos\psi \, d\psi = 0 \tag{5.52}$$

$$\int_0^{2\pi} \sin^2\psi \, d\psi = \pi$$

$$\int_0^{2\pi} \cos^2\psi \, d\psi = \pi$$

Breaking down (5.49), we obtain:

$$\frac{1}{2\pi}\int_0^{2\pi} \theta \cdot u_T^2 \, d\psi = \frac{1}{2\pi}\int_0^{2\pi} \theta_0 \cdot (x + \mu\sin\psi)^2 \, d\psi$$

$$= \theta_0 \cdot \left(x^2 + \frac{1}{2}\mu^2\right) \tag{5.53}$$

while:

$$\frac{1}{2\pi}\int_0^{2\pi} u_P \cdot u_T \, d\psi = \frac{1}{2\pi}\int_0^{2\pi} (\lambda + \beta\mu\cos\psi) \cdot (x + \mu\sin\psi) \, d\psi$$

$$= \lambda x \tag{5.54}$$

all other terms cancelling out after substituting for β and $(d\beta/d\psi)$ and integrating. Hence finally,

$$C_T = \int_0^1 sa\left[\theta_0\left(x^2 + \frac{1}{2}\mu^2\right) - \lambda x\right] dx$$

$$= sa\left[\frac{\theta_0}{3}\left(1 + \frac{3}{2}\mu^2\right) - \frac{\lambda}{2}\right] \tag{5.55}$$

This is the simplest expression for the lift coefficient of a rotor in forward level flight. The assumptions on which it is based are those assumed for hover in Chapter 3, namely uniform induced velocity across the disc, constant solidity s along the span and zero blade twist. As before, it may be assumed that for a linearly twisted blade, Equation 5.55 can be used if the value of θ is taken to be that at three-quarters radius. Also in Equation 5.55 the values of θ and λ are taken relative to the non-feathering plane as reference. Bramwell (p. 157) derives a significantly more complex expression for thrust when referred to disc axes (the TPP) but since the transformation involves the assumption that actual thrust, to the accuracy required, is not altered as between the two reference planes, the change is a purely formal one and Equation 5.55 stands as a working formula.

5.4.3 In-Plane H-force

In hover the in-plane H-force, representing principally the blade profile drag, contributed only to the torque. Here, however, since the resultant velocity at the blade is $\Omega r + V \sin \psi$ (Equation 5.35), the drag force on the advancing side exceeds the reverse drag force on the retreating side, leaving a net drag force on the blade, positive in the rearward direction.

Seen in azimuth (Figure 5.19) the elementary H-force, reckoned normal to the blade span and resolved in the rearward direction, is:

$$dH = (dD \cos \phi + dL \sin \phi) \sin \psi \tag{5.56}$$

which may be written as dH_0 plus dH_i, where the suffices relate to the profile drag and induced drag terms, respectively. Treating the drag term separately and making the usual approximations, we have:

$$dH_0 = dD \sin \psi$$
$$= \frac{1}{2} \rho U_T^2 \cdot c \, dr \cdot C_{D0} \cdot \sin \psi \tag{5.57}$$

In coefficient form, for N blades, this gives:

$$dC_{H_0} = \frac{\frac{1}{2} \rho N U_T^2 \cdot c \cdot C_{D0} \cdot \sin \psi \, dr}{\frac{1}{2} \rho (\Omega R)^2 \pi R^2} \tag{5.58}$$
$$= s \cdot u_T^2 \cdot C_{D0} \cdot \sin \psi \, dx$$
$$= s(x + \mu \sin \psi)^2 \cdot C_{D0} \cdot \sin \psi \, dx$$

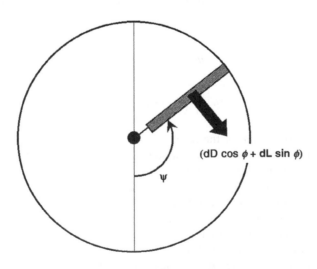

Figure 5.19 Elementary H-force

Hence:

$$C_{H_0} = s \cdot C_{D0} \int_0^1 \left[\frac{1}{2\pi} \int_0^{2\pi} (x + \mu \sin\psi)^2 \cdot \sin\psi \, d\psi \right] dx$$

$$= s \cdot C_{D0} \int_0^1 \mu x \, dx \qquad (5.59)$$

$$= \frac{1}{2} s \cdot C_{D0} \cdot \mu$$

Overall then for the in-plane H-force, we have:

$$C_H = \frac{1}{2} s C_{D0} \cdot \mu + C_{H_i} \qquad (5.60)$$

Expressions can be obtained for the induced component C_{H_i} in terms of θ, λ, μ and the flapping coefficients a_0, a_1 and b_1: these are derived in varying forms in the standard textbooks, for example Bramwell (p. 148) and Johnson (p. 177). The relations are somewhat complex and since we shall not require to make further use of them in the present treatment, and moreover in the usual case C_{H_i} is small compared with C_{H_0}, we can be satisfied with the reduction at Equation 5.60.

5.4.4 Torque and Power

The elementary torque is:

$$dQ = dH \cdot r$$
$$= r(dD\cos\phi + dL\sin\phi) \qquad (5.61)$$

Again there is a profile drag term, dQ_0 say, and an induced term dQ_i. The former is readily manipulated thus (in coefficient form):

$$C_{Q_0} = s \cdot C_{D0} \int_0^1 \left[\frac{1}{2\pi} \int_0^{2\pi} (x + \mu \sin\psi)^2 \cdot x \, d\psi \right] dx$$

$$= s \cdot C_{D0} \int_0^1 \left(x^3 + \frac{1}{2} \mu^2 x \right) dx \qquad (5.62)$$

$$= \frac{1}{4} s \cdot C_{D0} \cdot \left(1 + \mu^2 \right)$$

The induced term, after a lengthier manipulation, is shown (Bramwell, p. 151) to be:

$$C_{Q_i} = \lambda C_T - \mu C_{H_i} \qquad (5.63)$$

giving for the total torque:

$$C_Q = \frac{1}{4} s \cdot C_{D0} \cdot \left(1 + \mu^2 \right) + \lambda C_T - \mu C_{H_i} \qquad (5.64)$$

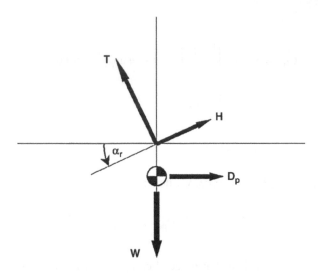

Figure 5.20   Forces in trimmed level flight

Using Equation 5.60 this becomes:

$$C_Q = \frac{1}{4}s \cdot C_{D0} \cdot (1+\mu^2) + \lambda C_T - \mu C_H + \frac{1}{2}s \cdot C_{D0} \cdot \mu^2$$

$$= \frac{1}{4}s \cdot C_{D0} \cdot (1+3\mu^2) + \lambda C_T - \mu C_H$$

(5.65)

Now by Equation 5.39 the inflow factor λ is a function of the inclination α_r of the TPP, which clearly depends upon the drag not only of the rotor but of the helicopter as a whole. Examining the relationships for trimmed level flight, illustrated in Figure 5.20, we have approximately:

$$T = W \tag{5.66}$$

$$T\alpha_r = H + D_P \tag{5.67}$$

D_P being the parasite drag of the fuselage, including tail rotor, tailplane and any other attachments.

Thus:

$$\alpha_r = \frac{H}{T} + \frac{D_P}{T}$$

$$= \frac{C_H}{C_T} + \frac{D_P}{W}$$

(5.68)

whence:

$$\lambda = \lambda_i + \mu\alpha_r$$

$$= \lambda_i + \mu\frac{C_H}{C_T} + \mu\frac{D_P}{W}$$

(5.69)

Using this in Equation 5.65, the power coefficient is expressed in the form:

$$C_P = C_Q = \lambda_i C_T + \frac{1}{4} s \cdot C_{D0} \left(1 + 3\mu^2\right) + \mu \frac{D_P}{W} C_T \tag{5.70}$$

which is seen to be the sum of terms representing the induced or lift-dependent drag, the rotor profile drag and the fuselage parasite drag. The first of these had already been derived (Equation 5.22) when considering the adaptation of momentum theory to forward flight.

In practice both the induced and profile drag power requirements are somewhat higher than are shown in Equation 5.70. An empirical correction factor k_i for the induced power was suggested in Equation 5.23. For the profile drag power the deficiency of the analytical formula arises from neglect of:

- a spanwise component of drag (Figure 5.19);
- a yawed-wing effect on the profile drag coefficient at azimuth angles significantly away from 90° and 270°;
- the reverse flow region on the retreating side.

The first of these factors is probably the most important. They are conventionally allowed for by replacing the factor 3 in Equation 5.70 by an empirical, larger factor, k say. Studies by Bennett [7] and Stepniewski [8] suggest that an appropriate value is between 4.5 and 4.7. Industrial practice tends to be based on a firm's own experience: thus a value commonly used by Westland Helicopters is 4.65.

With the empirical corrections embodied, the power equation takes the form:

$$C_P = k_i \lambda_i C_T + \frac{1}{4} s \cdot C_{D0} \left(1 + k\mu^2\right) + \mu \frac{D}{W} C_T \tag{5.71}$$

This will be followed up in the chapter on helicopter performance (Chapter 7). In the present chapter we take our analytical study of the rotor aerodynamics two stages further: firstly examining the nature of the flapping coefficients a_0, a_1 and b_1 in terms of θ, λ and μ; and secondly looking at some typical values of collective pitch θ, inflow factor λ and the flapping coefficients in relation to the forward speed parameter μ and the level of thrust coefficient C_T.

5.4.5   Flapping Coefficients

The flapping motion is determined by the condition that the net moment of forces acting on the blade about the flapping hinge is zero. Referring back to Figure 4.12, the forces on an element dr of blade span, of mass $m\,dr$, where m is the mass per unit span, are:

- the aerodynamic lift, expressed as an element of thrust dT, acting on a moment arm r;
- a centrifugal force $r\Omega^2 m\,dr$, acting on a moment arm $r\beta$;
- an inertial force $r\ddot{\beta}m\,dr$, acting on a moment arm r;
- a blade weight moment, small in comparison with the rest and therefore to be neglected.

These lead to the flapping moment relationship given in Equation 4.3. Writing the aerodynamic or thrust moment for the time being as M_T, we have:

$$\int_0^R \beta r^2 \Omega^2 m\,dr + \int_0^R \ddot{\beta} r^2 m\,dr = M_T \tag{5.72}$$

Assuming the spanwise mass distribution is uniform (i.e. m is constant), this equation integrates to:

$$\beta \frac{R^3}{3} \Omega^2 m + \ddot{\beta} \frac{R^3}{3} m = M_T \tag{5.73}$$

Substituting the first-order Fourier expressions for β and $\ddot{\beta}$ leads to:

$$\frac{R^3}{3} \Omega^2 m a_0 = M_T \tag{5.74}$$

Thus the aerodynamic moment M_T is invariant with azimuth angle ψ. If I is written for the moment of inertia of the blade about its hinge, that is to say:

$$I = \int_0^R mr^2 \, dr = \frac{1}{3} mR^3 \tag{5.75}$$

we have:

$$a_0 = \frac{M_T}{I\Omega^2} \tag{5.76}$$

Now M_T may be written:

$$M_T = \int_0^R r \frac{dT}{dr} \, dr = \frac{1}{2} \rho ac \int_0^R \left(\theta U_T^2 - U_P U_T \right) r \, dr \tag{5.77}$$

so that, in dimensionless form:

$$a_0 = \frac{1}{2} \gamma \int_0^R \left(\theta u_T^2 - u_P u_T \right) x \, dx \tag{5.78}$$

where γ is defined by:

$$\gamma = \frac{\rho ac R^4}{I} \tag{5.79}$$

and is known as the Lock number. It provides a ratio between the aerodynamic forces and the inertial forces which determine the centrifugal loads. Replacing u_T and u_P by their definitions in Equations 5.37 and 5.43, and substituting for β and $d\beta/d\psi$, the right-hand side of Equation 5.78 develops to:

$$\frac{1}{2} \gamma \int_0^1 \left[\theta \left(x^2 + \frac{1}{2} \mu^2 \right) - \lambda x + f_S(\sin\psi) + f_C(\cos\psi) \right] x \, dx \tag{5.80}$$

where f_S and f_C represent functions in $\sin\psi$ and $\cos\psi$ respectively.

Since M_T is independent of ψ, its value can be obtained by integrating only the first part of this expression. Hence:

$$a_0 = \frac{1}{2}\gamma \int_0^1 \left[\theta\left(x^2 + \frac{1}{2}\mu^2\right) - \lambda x\right] x \, dx$$

$$= \frac{1}{2}\gamma \int_0^1 \left[\theta\left(x^3 + \frac{1}{2}\mu^2 x\right) - \lambda x^2\right] dx \tag{5.81}$$

$$= \frac{1}{8}\gamma \left[\theta(1+\mu^2) - \frac{4\lambda}{3}\right]$$

This is for an untwisted blade ($\theta = $ constant θ_0) or in the usual way for a linearly twisted blade with θ taken at three-quarters radius.

Also, because of the independence of M_T the terms in $\sin\psi$ and those in $\cos\psi$ are each separately equatable to zero. These two equations yield expressions for the first harmonic coefficients a_1 and b_1, namely:

$$a_1 = \frac{\mu\left(\frac{8}{3}\theta_0 - 2\lambda\right)}{1 - \frac{1}{2}\mu^2} \tag{5.82}$$

$$b_1 = \frac{\frac{4}{3}\mu\theta_0}{1 + \frac{1}{2}\mu^2} \tag{5.83}$$

The three equations immediately above represent the classical definitions of flapping coefficients, in which θ and λ have been defined relative to the NFP. Equivalent, though rather more complex, definitions relative to the TPP are given by Johnson (p. 189) or Bramwell (p. 157). Bramwell's equations, while not completely general, are probably accurate enough for most purposes and are quoted here for ease of reference (*they are quoted in the order of calculation*):

$$a_1 = \frac{\mu\left[\frac{8}{3}\theta - 2\lambda_T\right]}{\left(1 + \frac{3}{2}\mu^2\right)} \tag{5.84}$$

$$a_0 = \frac{\gamma}{8}\left[\theta(1+\mu^2) - \frac{4}{3}(\lambda_T + \mu a_1)\right] \tag{5.85}$$

$$b_1 = \frac{\frac{4}{3}\mu a_0}{\left(1 + \frac{1}{2}\mu^2\right)} \tag{5.86}$$

The corresponding relationship for thrust coefficient is:

$$C_T = sa\left[\frac{\theta}{3}\left(1 + \frac{3}{2}\mu^2\right) - \frac{\lambda_T}{2} - \frac{\mu a_1}{2}\right] \tag{5.87}$$

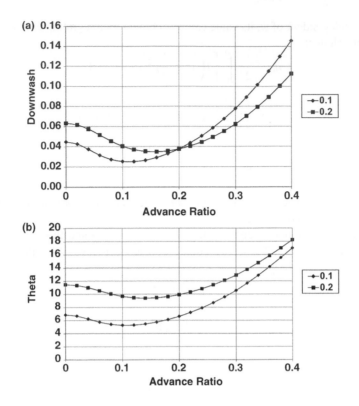

Figure 5.21   (a) Calculated values of λ_T v μ. (b) Collective pitch θ v μ

From the preceding discussion, two reference planes have been quoted, namely the TPP and the NFP. The previous analysis used the NFP; however, the rotor downwash is intimately linked with the rotor disc or the TPP. Bramwell's equations (5.84)–(5.87) quote the λ term in terms of the TPP. To avoid confusion, the subscript T is used for the downflow term (λ_T). This also links naturally with the definition of λ in Equation 5.39. The λ_T notation is also used in Figure 5.21.

5.4.6   Typical Numerical Values

Calculations have been made to illustrate in broad fashion the ways in which parameters discussed in the foregoing analysis vary with one another and particularly with forward speed. For this purpose the following values have been used:

Rotor solidity	s	0.08
Blade lift curve slope	a	5.7
Lock number	γ	8
Aircraft weight ratio	$\dfrac{W}{\frac{1}{2}\rho(\Omega R)^2 A}$	0.016
Parasite drag factor	$\dfrac{f}{A}$	0.016

The parasite drag factor is a form of expression in common use, in which f is the 'equivalent flat plate area' defined by:

$$D_P = \frac{1}{2}\rho V^2 f \tag{5.88}$$

D_P being the parasite drag and A the rotor disc area.

Figure 5.21a shows the variation of inflow factor λ with advance ratio μ at two levels of thrust coefficient. As previously mentioned, λ, as defined in Equation 5.39, is relative to the TPP, so is denoted by λ_T in the diagram. The variation shows a minimum value at moderate μ, inflow being high at low μ because the induced velocity is large and high again at high μ because of the increased forward tilt of the TPP required to overcome the parasite drag. The lower the thrust coefficient, the more marked the high-μ effect.

Figure 5.21b shows the corresponding variation of collective thrust angle θ, for $C_T/s = 0.2$. The variations of θ and λ are similar in character, as might be expected from Equation 5.87.

Combination of Equations 5.39 and 5.87 leads, on elimination of λ, to a direct relationship between C_T and θ which, using the chosen values of aircraft weight ratio and parasite drag factor in the final term, is:

$$B = \frac{1}{3}\left(1 + \frac{3}{2}\mu^2\right) + \mu^2\left(\frac{4}{3} - 2\mu^2\right) \tag{5.89}$$

where B is a slowly decreasing function of μ.

Note that when μ is zero, $B = \frac{1}{3}$ and $V_i/V_{i0} = 1$, so that we have Equation 3.29 as previously derived for the hover. Figure 5.22 shows variations of θ with C_T for different levels of μ. The characteristics at low and high forward speed are significantly different. When μ is zero or small the variation is nonlinear, θ increasing rapidly at low thrust coefficient owing to the induced flow term (the second expression in the equation) and more slowly at higher C_T as the first term becomes dominant. At high μ, however, the induced velocity factor V_i/V_{i0} is so small that the second term becomes negligible for all C_T, so the θ/C_T relationship is effectively linear. The intercept on the θ axis reflects the particular value of μ while, more interestingly,

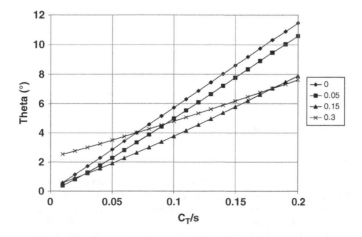

Figure 5.22 Variation of θ v C_T for values of μ

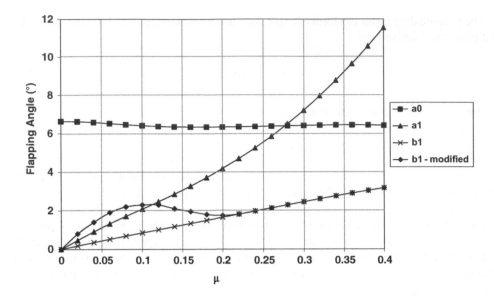

Figure 5.23   Variation of flapping coefficients v μ

with μ and s known the slope is a function only of the lift slope a. This provides an experimental method for determining a in a practical case.

A final illustration (Figure 5.23) shows the flapping coefficients a_0, a_1 and b_1 as functions of μ. These have been calculated using Equations 5.84–5.86. The coning angle a_0 varies only slightly with μ, being essentially determined by the thrust coefficient. It may readily be shown in fact that a_0 is approximately equal to $(3C_T\gamma)/(8sa)$ which with our chosen numbers has the value 0.105 rad or 6.0°. The longitudinal coefficient a_1 is approximately linear with forward speed, showing, however, an effect of the increase of λ at high speed. The lateral coefficient b_1 is also approximately linear, at about one-third the value of a_1. In practice b_1 at low speeds depends very much on the longitudinal distribution of induced velocity (assumed uniform throughout the calculations) and tends to rise to an early peak as indicated by a modified line in the diagram.

References

1. Glauert, H. (1926) A general theory of the autogiro, R & M, 1111.
2. Newman, S.J., Brown, R., Perry, J. *et al.* (2001) Comparitive numerical and experimental investigations of the vortex ring phenomenon in rotorcraft. 57th American Helicopter Society Annual National Forum, Washington, DC, 9–11 May, Vol. 2, pp. 1411–1430.
3. Drees, J. and Hendal, W. (1951) Airflow patterns in the neighbourhood of helicopter rotors. *J. Aircr. Eng.*, **23**, 107–1112.
4. Perry, F.J. (2000) Vortex ring instability in axial and forward flight – comparisons with test, Private Technical Note 0002/2000, June.
5. Brand, A., Kisor, R., Blyth, R. *et al.* (2004) V-22 high rate of descent (HROD) test procedures and long record analysis. 60th American Helicopter Society Annual National Forum, Baltimore, MD, June 7–10.
6. Mangler, K.W. and Squire, H.B. (1950) The induced velocity field of a rotor, R & M, 2642.
7. Bennett, J.A.J. (1940) Rotary wing aircraft. *Aircr. Eng. Aerosp. Technol.*, **12**, 139–146.
8. Stepniewski, W.Z. (1973) Basic aerodynamics and performance of the helicopter. *AGARD Lecture Series*, 63.

6

Aerodynamic Design

6.1 Introductory

In this chapter are described some of the trends in aerodynamic design which in the latter part of the twentieth century and the beginning of the twenty-first century are making the helicopter a considerably more efficient flying vehicle than it formerly was. In earlier years the low power-to-weight ratio of piston engines necessitated the use of large rotors to provide the all-important vertical lift capability: both profile drag and parasite drag were unavoidably high in consequence and forward speeds were therefore so low as to consign the problems of refining either the lift or drag performance to a low, even zero, priority. With the adoption of gas-turbine engines, and an ever-increasing list of useful and important applications for helicopters, in both military and civil fields of exploitation, forward flight performance has become a more lively issue, even to the point of encouraging comparisons with fixed-wing aircraft in certain specialized contexts (an example is given in Chapter 7). Some improvements in aerodynamics stem essentially and naturally from fixed-wing practice. A stage has now been reached at which these appear to be approaching and, in certain areas, to have arrived at, optimum levels in the helicopter application and therefore a substantial description here is appropriate. Further enhancements, concerned with the fundamental nature of the rotor system, may yet emerge to full development: one such is the use of higher harmonic control, which is described briefly. In the concluding section an account is given of a step-by-step method of defining the aerodynamic design parameters of a new rotor system.

6.2 Blade Section Design

In the design of rotor blade sections there is an a priori case for following the lead given by fixed-wing aircraft. It could be said, for instance, that the use of supercritical aerofoil sections for postponing the drag-rise Mach number is as valid an objective for the advancing blade of a rotor as for the wing of a high-subsonic transport aircraft. Or again, the use of blade camber to enhance maximum lift may be as valuable for the retreating blade as for a fixed wing approaching stall. Having accepted, say, this latter principle, there remains the problem of adapting it to the helicopter environment: this requires focused research, and substantial progress has been achieved.

The widely ranging conditions of incidence and Mach number experienced by a rotor blade in forward flight are conveniently illustrated by a 'figure-of-eight' diagram (sometimes called

Basic Helicopter Aerodynamics, Third Edition. John Seddon and Simon Newman.
© 2011 John Wiley & Sons, Ltd. Published 2011 by John Wiley & Sons, Ltd.

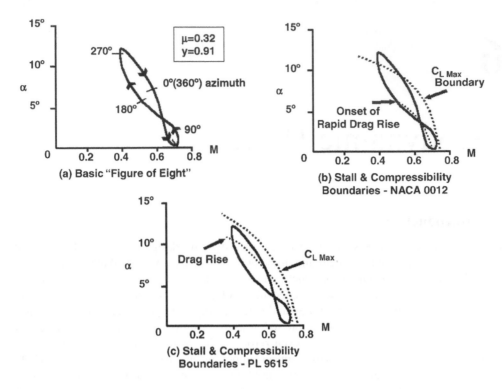

Figure 6.1   Figure-of-eight diagrams for a typical blade

sausage plots) (Figure 6.1a) which plots these conditions for a particular station on the blade near the tip ($x = 0.91$ in the case shown) at a specified value of μ. (In essence, a spanwise section of the blade is taken around a complete revolution and the various aerodynamic quantities recorded.) The hovering condition would be represented by a single point: as μ is increased the figure of eight expands, extending into regions of higher α (or C_L) and higher Mach number (M).

Plotting on such a diagram the α–M loci of $C_{L\,Max}$ and M_D (the drag-rise Mach number) for a particular blade section, these being obtained independently, as for example by two-dimensional section tests in a wind tunnel, will give an indication of whether either blade stall or drag divergence will be encountered in the rotor at the particular level of μ. Effectively, the figure-of-eight plot tracks a particular location of a rotor blade around one complete rotation of the rotor. This will highlight instances where the aerofoil section crosses over boundaries such as drag rise and stall. The example in Figure 6.1b relates to a symmetrical, 12% thick, NACA 0012 section. It is seen that the retreating-blade loop passes well into the stalled region and the advancing-blade loop likewise into the drag-rise region.

NACA 0012 was the standard choice for helicopter blade sections over many years. It is symmetric and is expressed mathematically. The function gives the section the following features:

- a parabolic nose shape of given radius;
- a specified thickness/chord ratio with maximum thickness at a specified chordwise location;
- the trailing edge has a given thickness and included angle.

Modern aerofoil sections embodying camber to increase maximum lift have been developed in various series, of which the 'VR' Series in the USA and the '96' Series in the UK are examples. Results for a 9615 section – the basis for the Westland Lynx blade – are shown in Figure 6.1c. The figure of eight now lies wholly within the $C_{L\,Max}$ locus, confirming an improvement in lift performance. Additionally the high-Mach-number drag rise now affects a much reduced portion of the retreating-blade loop, and the advancing-blade loop not at all, so a reduction in power requirement can be expected. It should be borne in mind that any improvement in thrust capability will automatically incur an increase in induced and climb power.

The evidence, though necessary, is not of itself sufficient, however. To ensure acceptability of the cambered section for the helicopter environment, additional aspects of a major character need to be considered. One is the question of section pitching moments. The use of camber introduces a nose-down C_{M0} (pitching moment at zero lift), which has an adverse effect on loads in the control system. A gain in $C_{L\,Max}$ must therefore be considered in conjunction with the amount of C_{M0} produced. One way of controlling the latter is by the use of reflex camber over the rear of a profile. Wilby [1] gives comparative results for a number of section shapes of the '96' Series, tested in a wind tunnel under two-dimensional steady-flow conditions.

A selection of his results appears in Figure 6.2, from which we can see that the more spectacular gains in $C_{L\,Max}$ (30–40%) tend to be associated with more adverse pitching

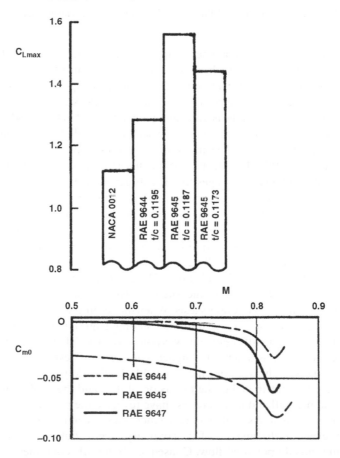

Figure 6.2 Comparison of $C_{L\,Max}$ and C_{M0} for various blade section profiles (after Wilby)

Figure 6.3   Spanwise variation of the aerofoil sections on the Merlin main rotor blade

moments, especially above a Mach number of approximately 0.75, which would apply on the advancing side of a rotor. Generally, therefore, compromises must be sought through much careful section shaping and testing. Moreover, while aiming to improve blade lift performance for the retreating sector, care must be taken to see that the profile drag is not increased, either at low C_L and high Mach number for the advancing sector, or at moderate C_L and moderate Mach number for the fore and aft sectors which in a balanced forward flight condition will carry the main thrust load. Figure 6.3 shows the variation of aerofoil sections on the BERP blade fitted to the fifth pre-prototype (PP5) EH101 aircraft. Each section is specifically designed for the particular incidence/Mach number ranges that it will experience.

While static testing of this nature is very useful in a comparative sense, it cannot be relied upon to give an accurate final value of $C_{L\ Max}$, because the stall of a rotor blade in action is known to be dynamic in character, owing to the changes in incidence occurring as the blade passes through the retreating sector. Farren [2] recorded, as long ago as 1935, that when an aerofoil is changing incidence, the stalling angle and $C_{L\ Max}$ may be different from those occurring under static conditions. Carta [3] in 1960 reported oscillation tests on a wing with 0012 section suggesting that this dynamic situation would apply in a helicopter context. Figure 6.4 shows a typical result of Carta's tests. When the aerofoil was oscillated through 6° on either side of 12° incidence (just above the static stalling angle), with a representative rotor frequency, a hysteresis loop in lift coefficient was obtained, in which the maximum C_L reached during incidence increase was about 30% higher than the static level.

Many subsequent researchers, among them Ham [4], McCroskey [5], Johnson and Ham [6] and Beddoes [7], have contributed to the provision of data and the evolution of theoretical treatments on dynamic stall and in the process have revealed the physical nature of the flow, which is of intrinsic interest. As blade incidence increases beyond the static stall point, flow reversals are observed in the upper surface boundary layer – but for a time these are not transmitted to the outside potential flow. Consequently the lift continues to increase with incidence. Eventually, flow separation develops at the leading edge (or it may be behind

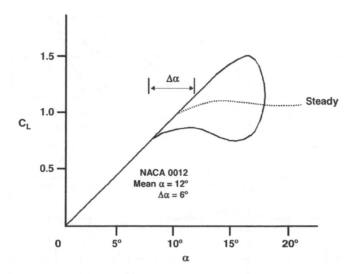

Figure 6.4   Lift hysteresis for oscillating blade (after Carta)

a recompression shock close to the leading edge), creating a transverse vortex which begins to travel downstream. As the vortex rolls back along the upper surface into the mid-chord region, lift continues to be generated but a large nose-down pitching moment develops owing to the redistribution of upper surface pressure. The passage of the vortex beyond the trailing edge results in a major breakdown of flow. Finally, when the incidence falls below the static stall angle as the blade approaches the rear of the disc, the flow reattaches at the leading edge and normal linear lift characteristics are re-established. The important consideration is that the aerodynamic influences with increasing incidence are not matched with the following reduction in incidence. This is the basis for the hysteretic effects seen in oscillating aerofoils.

Some further results for the RAE 9647 aerofoil section are shown in Figure 6.5, in this case from blade oscillation tests over four different incidence ranges. As the range is moved up the incidence scale, the hysteresis loop develops in normal force coefficient (representing C_L) and the pitching moment 'break' comes into play. In practice it is the latter which limits the rotor thrust, by reason of the large fluctuations in pitch control loads and in blade torsional vibrations which are triggered. It is of interest to note that in the results shown, the normal coefficient reached at the point of pitching moment break is about 1.8. Considerably higher values may in fact be attained; however, it is to be noted that this value on the retreating blade is not particularly important in itself, since what matters more is the amount of lift produced by the other blades in the fore and aft sectors, where in a balanced rotor the major contributions to thrust are made.

Initially, one saw a situation on blade section design still capable of further development. The emphasis was placed on improving the lift capability of the retreating blade. As the aspect of fuel economy in helicopter flight gains in importance, the incentive grows to reduce blade profile drag, particularly for the advancing sector. In this area there are probably improvements to be had by following the lead given by fixed-wing aircraft in the use of so-called supercritical wing sections. A further comment putting the incentive into context is made in Chapter 7. What is now under scrutiny is the influence of efficiency. This is a result of the perceived need to conserve energy resources and the aeroelastic behaviour of a blade is now being used to

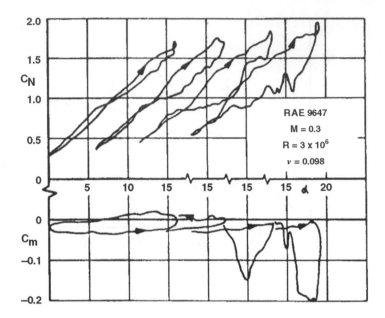

Figure 6.5 Development of lift hysteresis and pitching moment break as incidence range is raised (after Wilby)

improve performance and to reduce both fuel consumption and vibratory influences. The initial aeroelastic effects are passive in nature; however, the research direction is now directed towards active aerodynamic techniques such as blade morphing and tip blowing. This allows blade camber to be adjusted in a live environment and to control the nature and strength of the tip vortex.

6.3   Blade Tip Shapes

6.3.1   Rectangular

Performance of a rotor blade is governed, naturally, over its entire length. The tip region, however, plays a very influential role in the blade's aerodynamic character. For most of its formative years the helicopter rotor blade possessed a uniform blade chord out to the tip giving it a rectangular planform. A typical example is shown in Figure 6.6. The blade usually has a set of weights firmly fixed at the tip end for blade track and balance adjustments to be carried out. These are normally covered by a tip cap which covers these weights and restores an aerodynamically clean shape to the blade tip. For symmetric aerofoils, it is often a surface of revolution about the chordline.

6.3.2   Swept

The first move away from the rectangular blade planform was the inclusion of sweep in the tip region. It was abrupt and the sweep angle was constant. This is aimed at the advancing blade tip where high Mach numbers will be encountered at the higher forward speeds. The rearward

Figure 6.6 Rectangular blade planform of S61NM helicopter

movement of the blade chord at the tip end of the blade will require careful design as the lift centre for the tip and the local centre of gravity will now be behind the blade pitch axis and its shear centre. This will open up the possibility of blade flexing in pitch which can cause aeroelastic problems. An example of this type of tip planform is shown in Figure 6.7.

Figure 6.7 Simply swept rotor blade tip

Figure 6.8   Merlin main rotor blade tip (BERP)

6.3.3   Advanced Planforms

The loading on a helicopter blade is highly concentrated in the region of the tip, as has been seen (Figure 2.28). It is unlikely that a plain rectangular planform (a typical example is shown in Figure 6.6) is the optimum shape for the task of carrying this load and consequently investigations into tip design are a feature of modern aerodynamic research. Figure 6.8 shows the main rotor blade tip of the Merlin helicopter, which is the BERP planform with anhedral added (note the provision for picketing the blade and the leading edge erosion strip).

Figure 6.9 shows the main rotor blade tip of an Agusta Westland A129 variant. Since resultant velocities in the tip region on the advancing blade are close to Mach 1.0, it is natural to enquire whether sweepback can be incorporated to delay the compressibility drag rise and

Figure 6.9   A129 development main rotor blade tip

thereby reduce the power requirement at a given flight speed or alternatively raise the maximum speed attainable. The answer is not so immediately obvious as in the case of a fixed wing, because a rotor blade tip which at one moment is swept back relative to the resultant airflow, in the next moment lies across the stream. In fact, however, the gain from sweepback outweighs the loss, as is indicated in a typical case by Wilby and Philippe [8] (Figures 6.10a and b): a large reduction in Mach number normal to the leading edge is obtained over the rear half of the disc, including a reduction in maximum Mach number of the cycle (near $\psi = 90°$), at the expense of a small increase in the forward sector ($\psi = 130°$ to $240°$). Reductions in power required were confirmed in the case shown.

Shaping the blade tip can also be used to improve the stalling characteristics of the retreating blade. A particular all-round solution devised by Agusta Westland Helicopters is pictured in Figure 6.11. The principal features are:

- approximately $20°$ sweepback of the outboard 15% of blade span;

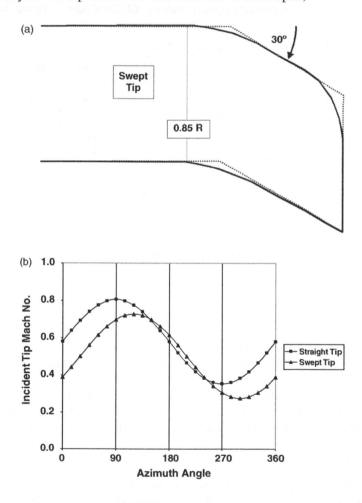

Figure 6.10 (a) Swept tip geometry. (b) Variation of Mach number normal to leading edge for straight and swept tips (after Wilby and Phillippe)

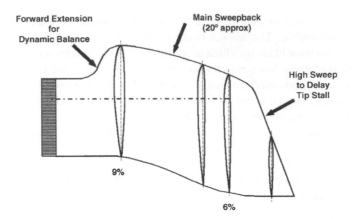

Figure 6.11   Agusta Westland development blade tip (BERP) (Courtesy Agusta Westland)

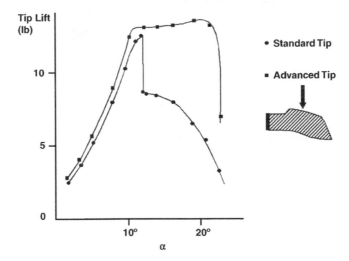

Figure 6.12   Wind tunnel results (non-oscillating) showing large advantage in stalling angle for Agusta Westland BERP tip (Courtesy Agusta Westland)

- a forward extension of the leading edge in this region, to safeguard dynamic stability;
- a sharply swept outer edge to promote controlled vortex separation and thereby delay the tip stall.

Wind tunnel tests (static conditions) showed this last effect to have been achieved in remarkable degree (Figure 6.12). Subsequently the tip proved highly successful in flight and was used on a version of the Lynx helicopter which captured the world speed record (see Chapter 7).

6.4   Tail Rotors[1]

The vast majority of helicopters are of the single main and tail rotor type. So far, any discussion has focused, primarily, on the main rotor. It provides most of the forces and moments required

---

[1] Some of the figures here are revisions of figures in *The Foundations of Helicopter Flight* by Simon Newman, Elsevier, 1994.

to fly successfully and under full control. However, it needs a mechanism for yaw control and this is where the tail rotor contributes. On a pure numbers basis, the main rotor provides control in 5 degrees of freedom:

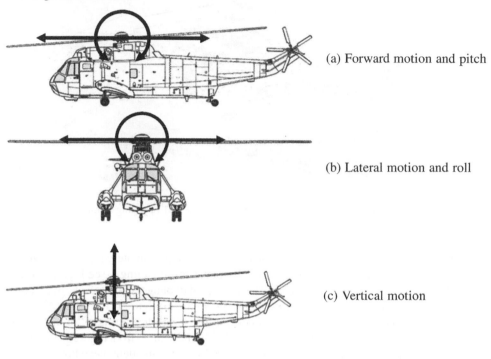

(a) Forward motion and pitch

(b) Lateral motion and roll

(c) Vertical motion

while the tail rotor provides the remaining sixth:

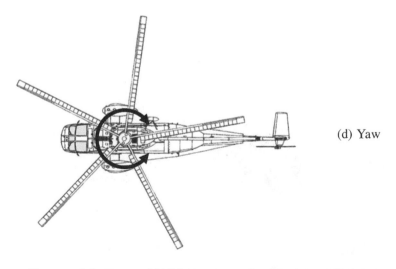

(d) Yaw

However, it has to provide this yaw control under very particular circumstances and in areas where the aerodynamic and dynamic forces on the blades are a real problem. It is positioned on a fin which is directly behind the main rotor hub and installation fairings. The tail rotor produces a thrust perpendicular to its disc plane which is reacted by a small lateral tilt of the

Figure 6.13 Puma hovering showing tail rotor thrust direction and opposing main rotor side force (Courtesy US Navy)

main rotor in opposition. It is placed at a vertical location at approximately the same height as the main rotor head in order to minimize any subsequent roll coupling – see Figure 6.13.

Its location at the rear extremity of the airframe can lead to dangers in the event of extreme manoeuvres as exemplified by the rapid deceleration shown in Figure 6.14.

The position of the tail rotor, behind the main rotor hub and top fuselage, results in the forward flight incident flow not being uniform over the tail rotor disc. Figure 6.15 shows a window placed in front of the tail rotor with an indication of the incident mean flow velocity. Efficiency is usually associated with uniformity and the tail rotor is no different. In order to account for the higher flow at the top of the tail rotor disc and the consequent lower flow at the

Figure 6.14 Extreme nose-up manoeuvre (Courtesy US Navy)

Figure 6.15   Incident flow over tail rotor (Courtesy US Navy)

bottom, a tail rotor rotational direction of backwards at the top would seem advisable. In most conventional helicopters, this is the case; however, there are exceptions. As will be discussed later, the early Westland Lynx helicopter had the opposite rotational direction which caused difficulties in side winds – in other words, there are differing reasons why this rotational direction has merits.

Because of the flight of a helicopter, a tail rotor has to operate in areas not experienced by the main rotor. The tail rotor has to overcome the effects of sideways flight (in either direction), autorotation and high-power climb. These conditions will require a pitch range of the order of 40°, see Byham [9]. This puts the blades at considerable pitch angles, which will introduce kinematic effects which are not normally experienced by the main rotor. One important effect is that of propeller moment.

6.4.1   Propeller Moment

With reference to Figure 6.16, consider a point of a blade at Q. It is positioned on a blade section defined by the plane PP'Q and with a zero-pitch blade has the coordinates x (chordwise) and y

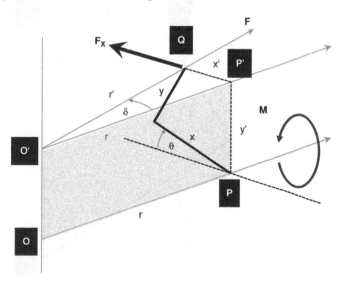

Figure 6.16

(thickness); however, the blade is rotated by an angle θ and the position of Q can be defined by the coordinates x' and y'.

Now Q will experience a centrifugal forcing term F in the direction shown in the figure. This will have a component, F_X, in the direction normal to the radial line through P, the origin of the blade rotation. This will exert a moment about the axis OP in the direction of opposing the blade pitch rotation with a moment arm, y'. This is the propeller moment and is given by:

$$M = F_X \cdot y' \tag{6.1}$$

The forces F and F_X are in a plane normal to the rotor shaft passing through the point Q.

The transformation from (x, y) to (x', y') can be seen using Figure 6.17.

We therefore have the relations:

$$\begin{aligned} x' &= x \cos \theta - y \sin \theta \\ y' &= x \sin \theta + y \cos \theta \end{aligned} \tag{6.2}$$

To evaluate the force component, F_X, Figure 6.18 shows the plane O'P'Q.

Now the centrifugal force F is given by:

$$F = \Omega^2 r' \, dm \tag{6.3}$$

where dm is the elemental mass of the blade at point Q.

Resolving gives:

$$F_X = F \sin \delta \tag{6.4}$$

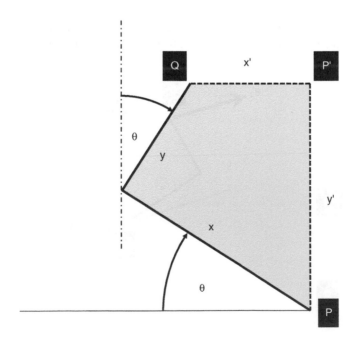

Figure 6.17

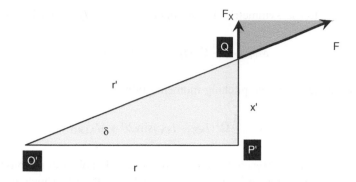

Figure 6.18

whence the force component is given by:

$$F_X = \Omega^2 r' \, dm \cdot \sin \delta$$
$$= \Omega^2 \, dm \cdot r' \sin \delta \qquad (6.5)$$
$$= \Omega^2 \, dm \cdot x'$$

and the elemental propeller moment now becomes:

$$dM = F_X \cdot y'$$
$$= \Omega^2 \, dm \cdot x' \cdot y'$$
$$= \Omega^2 \, dm (x \cos\theta - y \sin\theta)(x \sin\theta + y \cos\theta)$$
$$= \Omega^2 \, dm \left[(x^2 - y^2) \sin\theta \cos\theta + xy(\cos^2\theta - \sin^2\theta) \right] \qquad (6.6)$$
$$= \Omega^2 \, dm \left[(x^2 - y^2) \frac{\sin 2\theta}{2} + (xy) \cos 2\theta \right]$$

Finally, by integrating over the blade:

$$M_{\text{PROP}} = \int_{\text{BLADE}} \Omega^2 \left[\frac{1}{2} (x^2 - y^2) \sin 2\theta + xy(\cos 2\theta) \right] dm$$
$$= \Omega^2 \left[\frac{1}{2} (I_{XX} - I_{YY}) \sin 2\theta + I_{XY} \cos 2\theta \right]$$

$$I_{XX} = \int_{\text{BLADE}} x^2 \, dm \qquad (6.7)$$

$$I_{YY} = \int_{\text{BLADE}} y^2 \, dm$$

$$I_{XY} = \int_{\text{BLADE}} xy \, dm$$

If the tail rotor aerofoil is symmetric, the product of inertia, I_{XY}, vanishes, whence (6.7) simplifies to:

$$M_{\text{PROP}} = \frac{1}{2}\Omega^2(I_{XX} - I_{YY})\sin 2\theta \qquad (6.8)$$

To this is added the aerodynamic pitching moment giving:

$$M_{\text{TOTAL}} = \frac{1}{2}\Omega^2(I_{XX} - I_{YY})\sin 2\theta + M_{\text{AERO}} \qquad (6.9)$$

This pitching moment has to be reacted by the control system. For obvious reasons this has to be kept under limits of structural integrity and handling qualities – the pilot should not be expected to provide large forces on the pedals. The mechanism for adjusting the overall pitching moment is the bracketed term containing the difference of the two inertias. The moment can be increased by using I_{XX} and decreased using I_{YY}. Adjusting I_{XX} is relatively easy since the blade layout is predominantly in the chordwise direction; however, I_{YY} is not so easy as it can only be adjusted in the blade in the thickness direction, which is substantially lower because of the aerofoil section. To allow for this, the value of I_{YY} is varied by means of an external mass added to the blade construction. It is usually mounted on the blade cuff as shown in Figure 6.19.

These are termed *preponderance weights*.

Figure 6.19   Preponderance weight

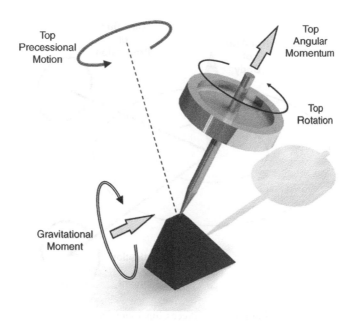

Figure 6.20   Precessing top

6.4.2   Precession – Yaw Agility

The phenomenon of gyroscopic precession is somewhat esoteric but it plays an important influence on tail rotor performance. To a first look, gyroscopic precession seems to disobey what is a natural understanding of the effect of moments on a body. The simple toy of a spinning top placed on a small tower seems to defy gravity. Instead of falling to the ground, it moves around a vertical axis only slowly leaning further and further, spiralling towards the ground plane on which the tower is standing. Figure 6.20 shows the top in question.

Using the right-hand rule, the top rotation and angular momentum are shown aligned with the spindle. The direction of the gravitational moment on the top is shown pointing to the right. The basic law is that the moment gives the rate of change of angular momentum. Hence the effect of gravity (and the reaction from the tower) will cause the angular momentum of the top to move in the same direction and thus the top rotates around the vertical axis in the precessional direction as shown. In Figure 6.21, the essential requirement is shown.

With reference to the rotational direction of the spinning body, the precessional rotation follows the applied moment by 90°.

The apparent avoidance of gravity is not a miracle, it just requires the moments and rotational directions to be expressed by vectors and then the standard laws apply – which they must!

The tail rotor will encounter precession principally in the spot turn manoeuvre. This is where the helicopter is in hover and, by applying an increase/decrease in tail rotor thrust, spins about an axis along the main rotor shaft. This is illustrated in Figure 6.22, where the aircraft is rotating nose-left.

The tail rotor disc will have to rotate about a vertical axis, even though it is rotating about the rotor shaft. In order for this to happen, the rotor must have an appropriate precessional moment applied to it – as described in Figure 6.23.

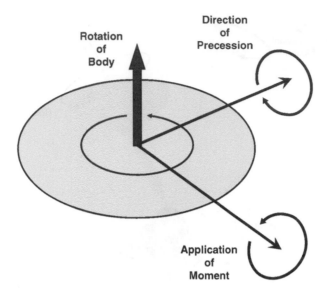

Figure 6.21   Precessional moment

This moment cannot be achieved by any other means than aerodynamic, since the blades are not rigidly attached to the rotor shaft. With a fixed collective pitch and no cyclic pitch provision, the only way a moment can be generated is by blade flapping. It is important to note that lift variation is dependent on blade flapping velocity and not flapping displacement. Bearing this in mind, Figure 6.24 shows a rotor, where there is a disc tilt caused by a flapping displacement. The flapping velocities can be seen to be 90° out of phase with the flapping displacement. We therefore have two phase angle changes to consider, which results in the rotation of a rotor about an axis in the plane of rotation as shown in Figure 6.25. The rotor disc flaps in a direction lagging behind the shaft movement – the two 90° phase angles sum to give a total phase change of 180°.

Figure 6.22   Spot turn manoeuvre

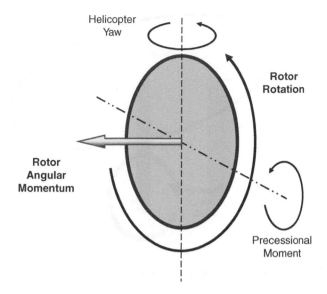

Figure 6.23 Tail rotor precessional moment

It can be shown that the rotation of a tail rotor about a vertical axis will generate a disc tilt of magnitude:

$$\text{Disc tilt} = \frac{16}{\gamma} \cdot \frac{q}{\Omega}$$

$$= \frac{16}{\gamma} \cdot \hat{q}$$

(6.10)

With the azimuth definition shown in Figure 6.26, a spot turn manoeuvre will cause a disc tilt which can be seen to be of a_1 variety.

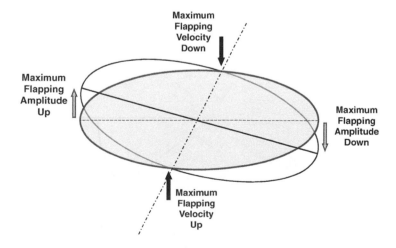

Figure 6.24 Relationship between flapping amplitude and flapping velocity

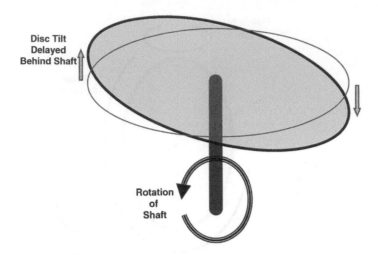

Figure 6.25   Disc tilt generated by gyroscopic precession due to rotating shaft

If we also consider the provision of pitch–flap coupling via a δ_3 hinge, the incidence of a typical blade is:

$$\alpha = \theta_0 - \frac{\dot{\beta}}{\Omega} - \beta \tan \delta_3 - \frac{V_Z + V_i}{\Omega r}$$

$$= \theta_0 - \beta' - \beta \tan \delta_3 - \frac{V_{ZD}}{\Omega r}$$

$$(6.11)$$

The values of V_Z and V_i are to be assumed constant over the rotor disc. Hence the V_Z term will represent the value at the centre of the rotor. The V_i value will be defined by actuator disc theory

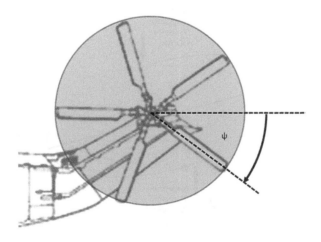

Figure 6.26   Azimuth definition

in climb at the appropriate value of V_Z. We define the flapping angle and velocity as:

$$\beta = a_0 - a_1 \cos \psi$$
$$\beta' = a_1 \sin \psi$$

(6.12)

from which we find from (6.11):

$$\alpha = \theta_0 - a_1 \sin \psi - (a_0 - a_1 \cos \psi) \tan \delta_3 - \frac{V_Z + V_i}{\Omega r}$$

$$= \theta_0 - a_0 \tan \delta_3 - a_1 (\sin \psi - \tan \delta_3 \cdot \cos \psi) - \frac{V_{ZD}}{\Omega r}$$

(6.13)

The a_1 term can be analysed by combining the trigonometric terms:

$$\sin \psi - \tan \delta_3 \cdot \cos \psi = k \sin(\psi - \phi)$$
$$= k \sin \psi \cos \phi - k \cos \psi \sin \phi$$

(6.14)

from which we must have:

$$k \sin \phi = \tan \delta_3$$
$$k \cos \phi - 1$$

(6.15)

giving:

$$\tan \phi = \tan \delta_3$$
$$\Rightarrow = \phi \delta_3$$
$$k = \sqrt{1 + \tan^2 \delta_3}$$
$$= \sec \delta_3$$

(6.16)

from which (6.13) becomes:

$$\alpha = \theta_0 - a_0 \tan \delta_3 - a_1 \sec \delta_3 \cdot \sin(\psi - \delta_3) - \frac{V_{ZD}}{\Omega r}$$

(6.17)

This will have an extreme value of:

$$\alpha = \theta_0 - a_0 \tan \delta_3 + a_1 \sec \delta_3 - \frac{V_{ZD}}{\Omega r}$$

(6.18)

If the blade is not to experience stall – taking the blade tip, as it has the highest Mach number – the collective pitch will be limited by this condition giving the stall incidence, α_S, that is:

$$\theta_{0\,\text{Max}} = \alpha_S + a_0 \tan \delta_3 - a_1 \sec \delta_3 + \frac{V_{ZD}}{\Omega R}$$

$$= \alpha_S + a_0 \tan \delta_3 - a_1 \sec \delta_3 + (\mu_Z + \lambda_i)$$

(6.19)

Now the rotor thrust will be given by:

$$T = \frac{1}{2} \rho V_{TT}^2 \cdot N_T c_T R_T \cdot a \cdot \frac{1}{2\pi} \int_0^{2\pi} d\psi \int_0^1 x^2 \left[\theta_0 - \beta' - \beta \tan \delta_3 - \frac{\mu_{ZD}}{x}\right] dx$$

(6.20)

which becomes:

$$\frac{C_{TT}}{sa} = \frac{1}{2\pi} \int_0^{2\pi} d\psi \int_0^1 x^2 \left[\theta_0 - \beta' - \beta \tan \delta_3 - \frac{\mu_{ZD}}{x} \right] dx$$

$$= \frac{1}{2\pi} \int_0^{2\pi} \left[\frac{\theta_0 - \beta' - \beta \tan \delta_3}{3} - \frac{\mu_{ZD}}{2} \right] d\psi \tag{6.21}$$

$$= \frac{1}{2\pi} \int_0^{2\pi} \left[\frac{\theta_0 - (a_1 \sin \psi) - (a_0 - a_1 \cos \psi) \tan \delta_3}{3} - \frac{\mu_{ZD}}{2} \right] d\psi$$

$$= \frac{\theta_0 - a_0 \tan \delta_3}{3} - \frac{\mu_{ZD}}{2}$$

Combining (6.19) and (6.21) gives the thrust maximum (without tip stall) as:

$$\frac{C_{TT\,Max}}{sa} = \frac{\theta_{0\,Max} - a_0 \tan \delta_3}{3} - \frac{\mu_{ZD}}{2}$$

$$= \frac{\alpha_S + a_0 \tan \delta_3 - a_1 \sec \delta_3 + \mu_{ZD} - a_0 \tan \delta_3}{3} - \frac{\mu_{ZD}}{2} \tag{6.22}$$

$$= \frac{\alpha_S - a_1 \sec \delta_3}{3} - \frac{\mu_{ZD}}{6}$$

Including the precessional result, we have finally:

$$\frac{C_{TT\,Max}}{sa} = \frac{\alpha_S}{3} - \frac{\mu_{ZD}}{6} - \frac{16}{3\gamma} \sec \delta_3 \cdot \frac{\dot{\Psi}}{\Omega_T}$$

$$\tag{6.23}$$

$$\frac{C_{TT\,Max}}{sa} = \frac{\alpha_S}{3} - \frac{\mu_Z}{6} - \frac{\lambda_i}{6} - \frac{16}{3\gamma} \sec \delta_3 \cdot \frac{\dot{\Psi}}{\Omega_T}$$

The thrust coefficient C_{TT} and the downwash λ_i are interrelated, which needs addressing.

6.4.3 Calculation of Downwash

The tail rotor is considered to be effectively in climb. To evaluate this we need to define the tail boom length, as shown in Figure 6.27.

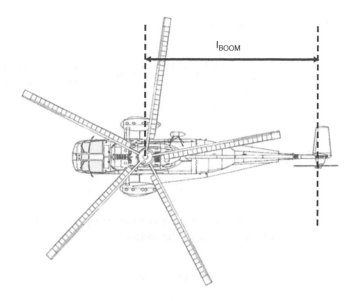

Figure 6.27   Definition of tail boom length

The normalized climb rate is given by:

$$\mu_Z = \frac{\dot{\Psi} \cdot l_{\text{BOOM}}}{V_{\text{TT}}}$$

$$= \frac{\dot{\Psi}}{\Omega_T} \cdot \frac{l_{\text{BOOM}}}{R_T} \tag{6.24}$$

Using momentum theory we obtain:

$$C_{\text{TT}} = 4\lambda_i(\mu_Z + \lambda_i) \tag{6.25}$$

For the limiting case where the extreme rotor blade tip is at the point of stall, we must have from (6.21) and (6.25):

$$C_{\text{TT Max}} = 4\lambda_{i\,\text{Max}}(\mu_Z + \lambda_{i\,\text{Max}})$$

$$\frac{C_{\text{TT Max}}}{sa} = \frac{\alpha_S}{3} - \frac{\mu_Z}{6} - \frac{\lambda_{i\,\text{Max}}}{6} - \frac{16}{3\gamma}\sec\delta_3 \cdot \frac{\dot{\Psi}}{\Omega_T} \tag{6.26}$$

These can be combined to give:

$$\lambda_{i\,\text{Max}}^2 + \left(\mu_Z + \frac{sa}{24}\right)\lambda_{i\,\text{Max}} - \frac{sa}{12}\left(\alpha_S - \frac{16}{\gamma} \cdot \frac{\dot{\Psi}}{\Omega_T} \cdot \sec\delta_3 - \frac{\mu_Z}{2}\right) = 0 \tag{6.27}$$

Introducing the result of (6.24) gives finally:

$$\lambda_{i\,Max}^2 + \left(\frac{\dot{\Psi}}{\Omega_T} \cdot \frac{l_{BOOM}}{R_T} + \frac{sa}{24} \right) \lambda_{i\,Max} - \frac{sa}{12} \left(\alpha_S - \frac{16}{\gamma} \cdot \frac{\dot{\Psi}}{\Omega_T} \cdot \sec\delta_3 - \frac{1}{2}\frac{\dot{\Psi}}{\Omega_T} \cdot \frac{l_{BOOM}}{R_T} \right) = 0$$

$$C_{TT\,Max} = 4\lambda_{i\,Max} \left(\frac{\dot{\Psi}}{\Omega_T} \cdot \frac{l_{BOOM}}{R_T} + \lambda_{i\,Max} \right)$$

(6.28)

6.4.4 Yaw Acceleration

Having established the maximum thrust that the tail rotor can generate, the maximum yaw acceleration can now be determined. We have the following:

$$\text{Yaw rate} = \dot{\Psi}$$
$$\text{Yaw accn} = \ddot{\Psi}$$

(6.29)

The yaw acceleration is proportional to the tail rotor thrust:

$$\ddot{\Psi} \propto T_T$$

(6.30)

Consider steady hover. The ideal induced power is given by:

$$P_{iM} = W\sqrt{\frac{W}{2\rho A}}$$

(6.31)

whence the total main rotor power can be determined via the figure of merit:

$$P_{TOT\,M} = W\sqrt{\frac{W}{2\rho A}} \cdot \frac{1}{FOM}$$

(6.32)

The main rotor torque is then:

$$Q_{TOT\,M} = \frac{W}{\Omega_M \cdot FOM}\sqrt{\frac{W}{2\rho A}}$$

$$= \frac{W}{\Omega_M \cdot FOM} \cdot \frac{V_{TM}}{2}\sqrt{C_{TM}}$$

$$= \frac{W}{2} \cdot \frac{R}{FOM} \cdot \sqrt{C_{TM}}$$

$$= T_{TH} \cdot l_{BOOM}$$

(6.33)

whence the hover tail rotor thrust can be expressed by:

$$T_{TH} = \frac{W}{2 \cdot FOM} \cdot \frac{R}{l_{BOOM}} \cdot \sqrt{C_{TM}}$$

$$= \frac{W}{2 \cdot FOM \cdot \overline{l_{BOOM}}} \cdot \sqrt{C_{TM}}$$

(6.34)

where:

$$\overline{l_{BOOM}} = \frac{l_{BOOM}}{R}$$

(6.35)

Returning to the yaw acceleration we have:

$$T_T = \frac{1}{2}\rho V_{TT}^2 \cdot A_T \cdot C_{TT}$$

(6.36)

whence:

$$\frac{T_T}{T_{TH}} = \frac{\frac{1}{2}\rho V_{TT}^2 \cdot A_T \cdot C_{TT}}{\frac{W}{2 \cdot FOM \cdot \overline{l_{BOOM}}} \cdot \sqrt{C_{TM}}}$$

$$= \frac{\frac{1}{2}\rho V_{TT}^2 \cdot A_T \cdot C_{TT} \cdot 2 \cdot FOM \cdot \overline{l_{BOOM}}}{W \cdot \sqrt{C_{TM}}}$$

(6.37)

Now, noting that:

$$W - \frac{1}{2}\rho V_{TM}^2 \cdot A \cdot C_{TM}$$

(6.38)

we have:

$$\frac{C_{TT}}{C_{TH}} = \frac{T_T}{T_{TH}} = \frac{\frac{1}{2}\rho V_{TT}^2 \cdot A_T \cdot C_{TT} \cdot 2 \cdot FOM \cdot \overline{l_{BOOM}}}{\frac{1}{2}\rho V_{TM}^2 \cdot A \cdot C_{TM} \cdot \sqrt{C_{TM}}}$$

$$= \left(\frac{V_{TT}}{V_{TM}}\right)^2 \cdot \left(\frac{R_T}{R}\right)^2 \cdot \left(2 \cdot FOM \cdot \overline{l_{BOOM}}\right) \cdot \frac{C_{TT}}{\left(C_{TM}\right)^{3/2}}$$

(6.39)

whence:

$$\frac{1}{C_{TH}} = \left(\frac{V_{TT}}{V_{TM}}\right)^2 \cdot \left(\frac{R_T}{R}\right)^2 \cdot \left(2 \cdot FOM \cdot \overline{l_{BOOM}}\right) \cdot \frac{1}{\left(C_{TM}\right)^{3/2}}$$

(6.40)

Hence, the yaw acceleration is given by the excess tail rotor thrust over the steady hover value. We can now examine the ratio of the excess tail rotor thrust at zero yaw rate and at a spot turn yaw rate giving the maximum yaw acceleration as:

$$\frac{\ddot{\Psi}_{\text{Max}}}{\ddot{\Psi}_{\text{H}}} = \frac{T_{\text{T Max}} - T_{\text{TH}}}{T_{\text{TH Max}} - T_{\text{TH}}}$$

$$= \frac{\dfrac{T_{\text{T Max}}}{T_{\text{TH}}} - 1}{\dfrac{T_{\text{TH Max}}}{T_{\text{TH}}} - 1}$$

$$= \frac{\dfrac{C_{\text{TT Max}}}{C_{\text{TH}}} - 1}{\dfrac{C_{\text{TTH Max}}}{C_{\text{TH}}} - 1}$$

$$= \frac{K \cdot C_{\text{TT Max}} - 1}{K \cdot C_{\text{TTH Max}} - 1}$$

(6.41)

where:

$$K = \left(\frac{V_{\text{TT}}}{V_{\text{TM}}}\right)^2 \cdot \left(\frac{R_{\text{T}}}{R}\right)^2 \cdot \left(2 \cdot \text{FOM} \cdot \overline{l_{\text{BOOM}}}\right) \cdot \frac{1}{\left(C_{\text{TM}}\right)^{3/2}}$$

(6.42)

6.4.5 Example – Sea King

The following data are used:

All-up weight, AUW	82 000	N
Main rotor tip speed, V_{TM}	206	m/s
Main rotor radius, R	9.39	m
Main rotor figure of merit, FOM	0.85	
Tail rotor tip speed, V_{TT}	207	m/s
Tail rotor radius, R_{T}	1.57	m
Tail rotor solidity, s	0.188	
Air density, ρ	1.2256	kg/m^3
Blade stall angle, α_{STALL}	12	degrees
Lift curve slope, a	5.8	/rad
Pitch–flap coupling angle, δ_3	45	degrees
Tail boom length, l_{BOOM}	11.2	m

This produces the graph in Figure 6.28.

The figure shows the yaw acceleration attainable with the rotor on the point of stall. The yaw acceleration is normalized with the value in steady hover – that is, no yaw rate. Notice that, as the yaw rate increases, the thrust potential reduces as the precessional effects emerge by limiting the collective pitch which can be achieved before part of the tail rotor enters stall – in this case a yaw rate of 2.2 rad/s.

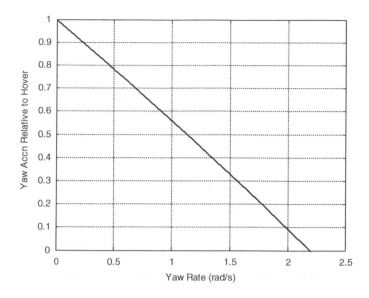

Figure 6.28   Yaw acceleration limit due to precession caused by yaw rate

6.5   Parasite Drag

Parasite drag – the drag of the many parts of a helicopter, such as the fuselage, rotor head, landing gear, tail rotor and tail surfaces, which make no direct contribution to main rotor lift – becomes a dominant factor in aircraft performance at the upper end of the forward speed range. Clearly the incentive to reduce parasite drag grows as emphasis is placed on speed achievement or on fuel economy. Equally clearly, since the contributing items all have individual functions of a practical nature, their design tends to be governed by practical considerations rather than by aerodynamic desiderata. Recommendations for streamlining, taken on their own, tend to have a somewhat hollow ring. What the research aerodynamicist can and must do, however, is provide an adequate background of reliable information which allows a designer to calculate and understand the items of parasite drag as they relate to their particular requirement and so review their options.

Such a background has been accumulated through the years and much of what is required can be obtained from review papers, of which an excellent example is that of Keys and Wiesner [10]. These authors have provided, by means of experimental data presented non-dimensionally, values of fuselage shape parameters that serve as targets for good aerodynamic design. These include such items as corner radii of the fuselage nose section, fuselage cross-section shape, afterbody taper and fuselage camber. Guidelines are given for calculating the drag of engine nacelles and protuberances such as aerials, lights and handholds. Particular attention is paid to the trends of landing-gear drag for wheels or skids, exposed or faired. Obviously the best solution for reducing the drag of landing gear is full retraction, which, however, adds significantly to aircraft weight. Keys and Wiesner have put this problem into perspective by means of a specimen calculation, which for a given mission estimates the minimum flight speed above which retraction shows a net benefit. The longer the mission, the lower the break-even speed.

The largest single item in parasite drag is normally the contribution from the rotor head, known also briefly as hub drag. This relates to the driving mechanism between rotor shaft and

Figure 6.29   Main rotor head of Merlin helicopter

blades, illustrated in Figure 6.29, and includes as drag components the hub itself, the shanks linking hub to blades, the hinge and feathering mechanisms and the control rods. Conventionally all these components are non-streamlined parts creating large regions of separated flow and giving a total drag greater than that of the basic fuselage, despite their much smaller dimensions. The drag of an articulated head may amount to 40% or 50% of total parasite drag, that of a hingeless head to about 30%. The application of aerodynamic fairings is possible to a degree, the more so with hingeless than with articulated heads, but is limited by the relative motions required between parts.

Sheehy [11] conducted a review of drag data on rotor heads from US sources and showed that projected frontal area was the determining factor for unfaired heads. Additionally, allowance had to be made for the effects of local dynamic pressure and head–fuselage interference, both of which factors increased the drag. Fairings needed to be aerodynamically sealed, especially at the head–fuselage junction. The effect of head rotation on drag was negligible for unfaired heads and variable for faired heads. The determination of helicopter hub drag is a very complex process and is often achieved by wind tunnel testing. A typical installation is shown in Figure 6.30.

Picking up the lines of Sheehy's review, a systematic series of wind tunnel model tests was made at Bristol University [12], in which a simulated rotor head was built up in stages. The drag results are summarized in Figure 6.31.

An expression for rotor head drag D emerges in the form:

$$\frac{D}{q_0} = \frac{q}{q_0} C_D A_P \left(1 - \frac{A_Z}{A_P} + \frac{A_S}{A_P} \right) \tag{6.43}$$

with the following definitions.

q_0 is the free stream dynamic pressure $\frac{1}{2}\rho V_0^2$. q is the local dynamic pressure at the hub position, measured in absence of the rotor head. In a general case, the local supervelocity and hence q can be calculated from a knowledge of the fuselage shape.

C_D is the effective drag coefficient of the bluff shapes making up the head. This may be assumed to be the same as for a circular cylinder at the same mean Reynolds number. For the

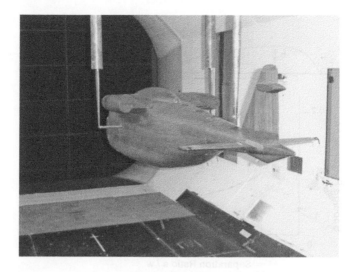

Figure 6.30   Pathfinder fuselage model in University of Southampton 11 ft by 8 ft wind tunnel

results of Figure 6.31 it is seen that a value $C_D = 1.0$ fits the experimental data well, apart from an analytically interesting but unreal case of the hub without shanks, where the higher Reynolds number of the large-diameter unit is reflected in a lower C_D value. In default of more precise information it is suggested that the value $C_D = 1.0$ should be used for general estimation purposes. One might expect the larger Reynolds number of a full-scale head to give a lower drag coefficient, but the suggestion rests to a degree on Sheehy's comment that small-scale model tests tend to undervalue the full-scale drag, probably because of difficulties of accurately modelling the head details.

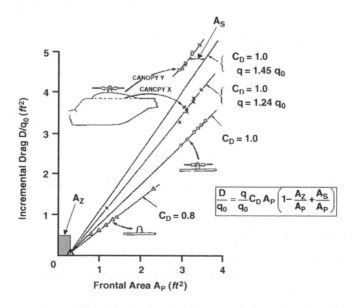

Figure 6.31   Rotor head drag results

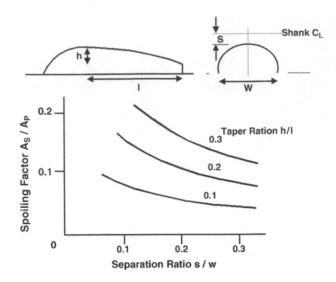

Figure 6.32   Chart for estimating spoiling factor A_S

A_P is the projected frontal area of the head, as used by Sheehy. A_Z represents a relieving factor on the drag, illustrated in Figure 6.31 and resulting from the fact that the head is partly immersed in the fuselage boundary layer. In magnitude A_Z turns out to be equal to the projected area contained in a single thickness of the boundary layer as estimated in absence of the head. The last quantity A_S represents in equivalent area terms the flow spoiling effect of the head on the canopy. This is a function jointly of the separation distance of the blade shanks above the canopy (the smaller the separation, the greater the spoiling) and the taper ratio of the canopy afterbody (the sharper the taper, the greater the spoiling). The ratio A_S/A_P may be estimated from the chart given in Figure 6.32 constructed by interpolation from the results for different canopies tested.

In light of the evidence quoted, the situation on rotor head drag may be summed up in the following points:

- The high drag of unfaired rotor heads is explained in terms of exposed frontal area and interference effects and can be calculated approximately for a given case.
- Hingeless systems have significantly lower rotor head drag than articulated systems.
- The scope for aerodynamic fairings is limited by the mechanical nature of the systems but some fairings are practical, more especially with hingeless rotors, and can give useful drag reductions.
- The development of head design concepts having smaller exposed frontal areas carries considerable aerodynamic benefit.

6.6   Rear Fuselage Upsweep

A special drag problem relates to the design of the rear fuselage upsweep for a helicopter with rear loading doors, where the width across the back of the fuselage needs to be more or less constant from bottom to top.

Figure 6.33 shows the difference in rear fuselage shape for a Merlin prototype in flight and a development (RAF) with a rear loading ramp door. In the 1960s, experience on fixed-wing

Figure 6.33 Shapes of Merlin fuselage both standard and that fitted with a rear loading ramp

aircraft [13] revealed that where a rear fuselage was particularly bluff, drag was difficult to predict and could be considerably greater than would have been expected on a basis of classical bluff-body flow separation. Light was thrown on this problem in the 1970s by T. Morel [14, 15]. Studying the drag of hatchback automobiles he found that the flow over a slanted base could take either of two forms: (1) the classical bluff-body flow consisting of cross-stream eddies or (2) a flow characterized by streamwise vortices. Subsequently the problem was put into a helicopter context by Seddon [16], using wind tunnel model tests of which the results are summarized in Figures 6.34–6.37.

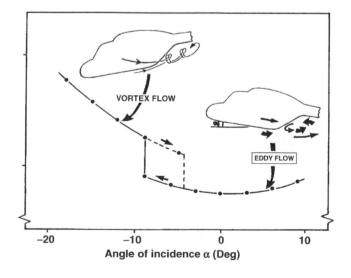

Figure 6.34 Types of flow from rear fuselage upsweep with associated critical drag change

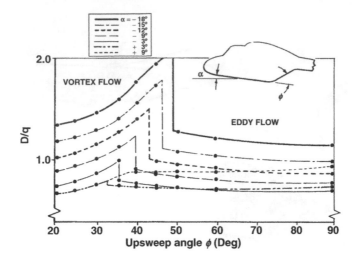

Figure 6.35   Variation of drag with upsweep angle at constant incidence

The combination of upsweep angle of the rear fuselage and incidence of the helicopter to the air stream determines the type of flow obtained. At positive incidence eddy flow persists. As incidence is decreased (nose going down as in forward flight) a critical angle is reached at which the flow changes suddenly to the vortex type and the drag jumps to a much higher level (Figure 6.34), which is maintained for further incidence decrease. If incidence is now increased, the reverse change takes place, though at a less negative incidence than before. The high drag corresponds to a high level of suction on the inclined surface, which is characteristic of the vortex flow. The suction force also has a downward lift component which is additionally detrimental to the helicopter. The type of flow is similar to that found on aerodynamically slender wings (as for example on the supersonic Concorde aircraft) but there the results are favourable because the lift component is upwards and the drag component is small except at high angle of attack.

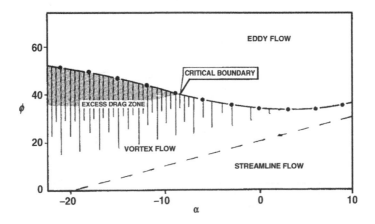

Figure 6.36 α–ϕ diagram showing all types of flow and indicating excess drag region

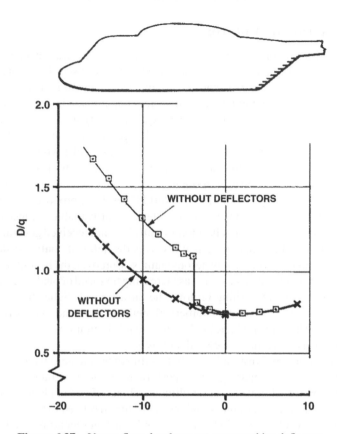

Figure 6.37   Vortex flow development prevented by deflectors

The effect of changing upsweep angle is shown in Figure 6.35. Here each curve is for a constant fuselage incidence. With upsweep angles near 90°, eddy flow exists as would be expected. At a point in the mid-angle range of upsweep, depending on incidence, the flow change occurs, accompanied by the drag increase. As the upsweep angle is further reduced the drag falls progressively but there is a significant range of angle over which the drag is higher than in eddy flow.

As an aid to design, the situation can be presented in the form of an α–ϕ diagram, ϕ being here the upsweep angle. The full line in Figure 6.36 is the locus of the drag jump when incidence is decreasing. If required, a locus can be drawn alongside to represent the situation with incidence increasing. Below the critical boundary is the zone of excess drag. From such a diagram, a designer can decide what range of upsweep angles is to be avoided for the aircraft. Of associated interest is the broken line shown: this marks an estimated boundary between vortex flow and streamlined flow, that is when no separation occurs at the upsweep. General considerations of aerodynamic streamlining suggest that the flow will remain attached if the upswept surface is inclined at not more than 20° to the direction of flight, in other words when $\phi - \alpha < 20°$.

The final diagram, Figure 6.37, shows that if vortex flow occurs naturally, it can be prevented by an application of short, closely spaced deflectors on the fuselage side immediately ahead of the upswept face. The action is one of preventing the vortex from building up by cutting it off at multiple points along the edge.

6.7 Higher Harmonic Control

In forward flight with a rotor operating under first-order cyclic control, a considerable proportion of the lifting capacity of the blades has been sacrificed, as we have seen in Chapter 4, in order to balance out the roll tendency. The lift carried in the advancing sector is reduced to very low level, while the main load is taken in the fore and aft sectors but at blade incidences (and hence lift coefficients) well below the stall. This can be seen explicitly in a typical figure-of-eight diagram, for example that in Figure 6.1c. Little can be done to change the situation in the advancing sector, but in the fore and aft sectors, where the loading has only a minor effect on the roll problem, the prospect exists of producing more lift without exceeding stalling limits in the retreating sector. In principle the result can be achieved by introducing second and possibly other harmonics into the cyclic control law. The concept is not new: Stewart [17] in 1952 proposed the use of second-harmonic pitch control, predicting an increase of at least 0.1 in available advance ratio. Until the 1980s, however, the potential of higher harmonic control has not received general development. Overall the problem is not a simple one, as it involves the fields of control systems and rotor dynamics at least as extensively as that of aerodynamics. Moreover, the benefits could until now be obtained by less complicated means, such as increasing tip speed or blade area. As these other methods reach a stage of diminishing returns, the attraction of higher harmonic control is enhanced by comparison. Also, modern numerical methods allow the rotor performance to be related to details of the flow and realistic blade aerodynamic limitations, so that the prediction of performance benefits is much more secure than it was.

A calculation provided by Westland Helicopters illustrates the aerodynamic situation. The investigation consisted in comparing thrust performances of two rotors with and without second-harmonic control, of quite small amplitude – about 1.5° of blade incidence. Local lift conditions near the tip were monitored round the azimuth and related to the C_L–M boundary of the blade section. The results shown in Figure 6.38 indicate that second-harmonic control gave an advantage of at least 0.2 in lift coefficient in the middle Mach number region appropriate to the disc fore and aft sectors.

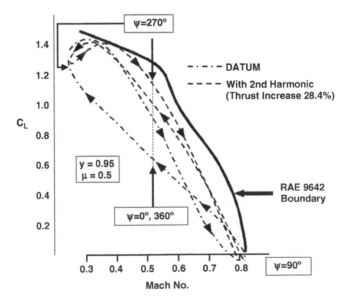

Figure 6.38   Use of second-harmonic control: calculated example

This translated into a 28.4% increase in thrust available for the same retreating-blade boundary. A further advantage was that the rotor with second-harmonic control required a 22% smaller blade area than the data rotor, which, whether exploited as a reduction say from six blades to five or as a weight saving at equal blade numbers, would represent a considerable benefit in terms of component size and mission effectiveness.

6.8 Aerodynamic Design Process

To complete this chapter we turn from research topics to the practical problem of determining the aerodynamic design of a rotor for a new helicopter project as shown in Figure 6.39.

A step-by-step process enables the designer to take into account the many and varied factors that influence their choice – aircraft specification, limitations in hover and at high forward speed, engine characteristics at various ratings, vibratory loads, flyover noise and so on. The following exposition comes from an unpublished instructional document kindly supplied by Westland Helicopters.

The basic requirement is assumed to be for a helicopter of moderate size, payload and range, with good manoeuvrability, robustness and reliability. The maximum flight speed is to be at least 80 m/s and a good high-temperature altitude performance is required, stipulated as 1200 m at ISA + 28 K. Prior to determining the rotor configuration, a general study of payload and range diagrams, in relation to the intended roles, leads to a choice of all-up weight, namely 4100 kg. Empty weight is set at 55% of this value, leaving 45% disposable weight, of which it is assumed one-half can be devoted to fuel and crew. Consideration of various engine options follows and a choice is made of a pair of engines having a continuous power rating at sea level ISA of 560 kW each, with take-off and contingency ratings to match. Experience naturally plays a large part in the making of these choices, as indeed it does throughout the design process.

First choice for the rotor is the tip speed: this is influenced by the factors shown in Figure 6.40. The tip Mach number in hover is one possible limitation. Allowing a margin for the fact that in high-speed forward flight a blade at the front or rear of the disc will be close to the same Mach number as in hover but at a higher lift coefficient, corresponding to the greater power required, the hover tip speed limit is set at Mach 0.69 (235 m/s). On the advancing tip in forward flight the lift coefficient is low and the Mach number limit can be between 0.8 and 0.9; recognizing that an advanced blade section will be used, the limit is set at 0.88. Flyover noise is largely a function of advancing tip Mach number and may come into this consideration. High

Figure 6.39 Typical helicopter design constraints

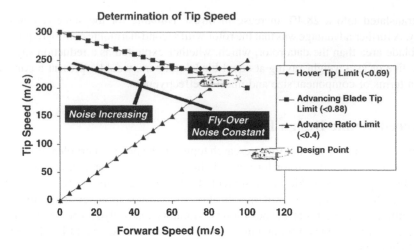

Figure 6.40   Determination of rotor tip speed for new rotor design

advance ratio brings on rotor vibratory loads and hence fuselage vibration, so a limiting μ for normal maximum speed is set at 0.4. Lastly the maximum speed specified is at least 80 m/s. It is seen that the satisfaction of these requirements constrains the rotor tip speed to about 215 m/s, the targeted maximum flight speed being 160 knots (82 m/s).

Next to be decided is the blade area. The area required increases as design speed increases, because the retreating blade operates at decreasing relative speed while its lift coefficient is stall limited. The non-dimensional thrust coefficient C_T/s is limited as shown in Figure 6.41a – see Equation 3.45. Writing:

$$\frac{C_T}{s} = \frac{W}{\frac{1}{2}\rho A(\Omega R)^2} \cdot \frac{\pi R}{Nc}$$

$$= \frac{W}{\frac{1}{2}\rho NcR(\Omega R)^2} \tag{6.44}$$

we have for the total blade area NcR:

$$NcR = \frac{2W}{\rho(\Omega R)^2} \bigg/ \frac{C_T}{s} \tag{6.45}$$

From a knowledge of tip speed (ΩR) and aircraft weight the blade area diagram, Figure 6.41b, is constructed. The design maximum speed then corresponds to a total blade area of 10 m^2. Note that use of the advanced blade section results in about 10% saving in blade area, which translates directly into rotor overall weight.

Choice of the rotor radius requires a study of engine performance. For the vertical axis in Figure 6.42, specific power loading (kW/kg) from the engine data is translated into actual power in W for the 4100 kg helicopter. Both twin-engine and single-engine values are shown, in each case for take-off, continuous and contingency ratings. Curves of power required for various hover conditions are plotted in terms of disc loading (kg/m^2) on the established basis

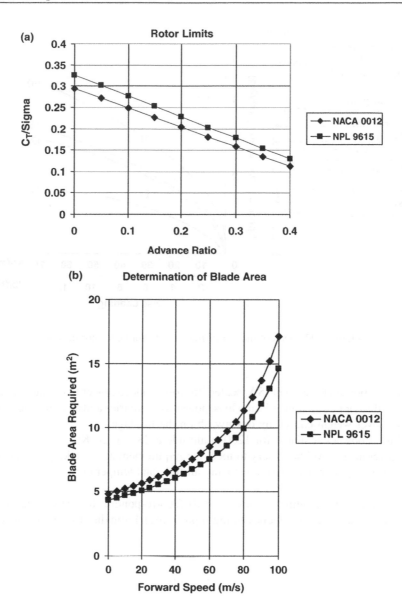

Figure 6.41 (a) Rotor limits for new rotor design. (b) Determination of blade area for new rotor design

(Chapter 2) that induced power is proportional to the square root of disc loading. The four curves shown, reading from the lowest upwards, are:

(1) ideal induced power at sea level ISA, given by:

$$P_i = W\sqrt{\frac{\omega}{2\rho}} \tag{6.46}$$

ω being the disc loading;

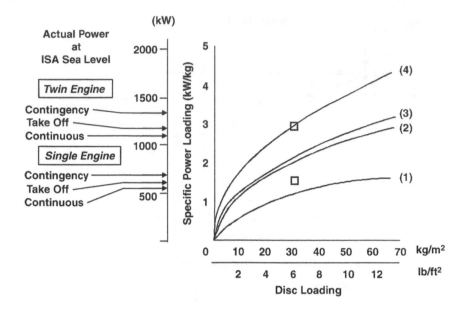

Figure 6.42 Determination of rotor radius for new rotor design

(2) actual total power at sea-level ISA, scaled up from induced power to include blade profile power, tail rotor power, transmission loss, power to auxiliaries and an allowance for excess of thrust over weight caused by downwash on the fuselage;
(3) actual total power calculated for 1200 m altitude at ISA + 28°K;
(4) total power at sea level necessary to meet the requirement at (3), taking into account the decrease of engine power with increasing altitude and temperature.

A design point for disc loading can now be read off corresponding to the twin-engine take-off power rating (or using the contingency rating if preferred). From disc loading the blade radius follows, since:

$$\omega = \frac{W}{A} = \frac{W}{\pi R^2} \tag{6.47}$$

Hence:

$$R = \sqrt{\frac{W}{\pi \omega}} \tag{6.48}$$

In the present example the selected disc loading is 32 kg/m^2 (314 N/m^2) and the corresponding blade radius is 6.4 m. The single-engine capability has also to be considered. It is seen that on contingency rating the helicopter does not have quite enough power from a single engine to hover at sea-level ISA and full all-up weight. The deficit is small enough, however, to ensure that a good fly-away manoeuvre would be possible following an engine failure; while at 90% all-up weight, hovering at the single-engine contingency rating is just possible.

Undetermined so far is the number of blades. From a knowledge of the blade radius and total blade area, the blade aspect ratio is given by:

$$\frac{R}{c} = \frac{R^2 N}{NcR} = 4.1N \qquad (6.49)$$

Using three blades, an aspect ratio of 12.3 could be considered low from a standpoint of three-dimensional effects at the tip. Five blades, giving aspect ratio 20.5, could pose problems in structural integrity and in complexity of the rotor hub and controls. Four blades are therefore the natural choice. Consideration of vibration characteristics is also important here. Vibration levels with three blades will tend to be high and with a reasonable flap–hinge offset, the pitch and roll vibratory moments (at $N\Omega$ frequency) will be greater for four blades than five. This illustrates that while a four-bladed rotor is probably the choice, not all features are optimum.

The choice between an articulated and a hingeless rotor is mainly a matter of dynamics and relates to flight handling criteria for the aircraft. A criterion often used is a time constant in pitch or roll when hovering; this is the time required to reach a certain percentage – 60% or over – of the final pitch or roll rate following an application of cyclic control. For the case in point, recalling the requirement for good manoeuvrability, low time constants are targeted. It is then found that, using flapping hinges with about 4% offset, the targets cannot be reached except by mounting the rotor on a very tall shaft, which is incompatible with the stated aims for robustness and compactness. A hingeless rotor produces greater hub moments, equivalent to flapping offsets of 10% and more, and is therefore seen as the natural choice.

References

1. Wilby, P.G. (1980) The aerodynamic characteristics of some new RAE blade sections and their potential influence on rotor performance. *Vertica*, **4**, 121–133.
2. Farren, W.S. (1935) Reaction on a wing whose angle of incidence is changing rapidly, R & M, 1648.
3. Carta, F.O. (1960) Experimental investigation of the unsteady aerodynamic characteristics of NACA 0012 airfoil, United Aircraft Laboratory Report, M-1283-1.
4. Ham, N.D. (1968) Aerodynamic loading on a two-dimensional airfoil during dynamic stall. *AIAA J.*, **6** (10), 1927–1934.
5. McCroskey, W.J. (1972) Dynamic stall of airfoils and helicopter rotors. AGARD Report 595.
6. Johnson, W. and Ham, N.D. (1972) On the mechanism of dynamic stall. *JAHS*, **17** (4), 36–45.
7. Beddoes, T.S. (1976) A synthesis of unsteady aerodynamic effects including stall hysteresis. *Vertica*, **1** (2), 113–123.
8. Wilby, P.G. and Philippe, J.J. (1982) An investigation of the aerodynamics of an RAE swept tip using a model rotor. Eighth European Rotorcraft Forum, Aix-en-Provence, Paper 25.
9. Byham, G.M. (1990) An overview of conventional tail rotors, in *Helicopter Yaw Control Concepts*, The Royal Aeronautical Society.
10. Keys, C.R. and Wiesner, R. (1975) Guidelines for reducing helicopter parasite drag. *JAHS*, **20**, 31–40.
11. Sheehy, T.W. (1975) A general review of helicopter hub drag data. Paper for Stratford AHS Chapter Meeting.
12. Seddon, J. (1979) An analysis of helicopter rotorhead drag based on new experiment. Fifth European Rotorcraft Forum, Amsterdam, Paper 19.
13. Lowe, B.G. and Trebble, W.J.G. (1968) Drag analysis on the Short SC5 Belfast, RAE Report.
14. Morel, T. (1976) The effect of base slant on the flow pattern and drag of 3-D bodies with blunt ends. General Motors Research Laboratories Symposium.
15. Morel, T. (1978) Aerodynamic drag of bluff body shapes characteristic of hatchback cars. SAE Congress and Exposition, Detroit, MI.
16. Seddon, J. (1982) Aerodynamics of the helicopter rear fuselage upsweep. Eighth European Rotorcraft Forum, Aix-en-Provence, Paper 2.12.
17. Stewart, W. (1952) Second harmonic control in the helicopter rotor, ARC R&M 2997, London.

7

Performance

7.1 Introduction

The preceding chapters have been mostly concerned with establishing the aerodynamic characteristics of the helicopter main rotor. We turn now to considerations of the helicopter as a total vehicle. The assessment of helicopter performance, like that of a fixed-wing aircraft, is at bottom a matter of comparing the power required with that available, in order to determine whether a particular flight task is feasible. The number of different performance calculations that can be made for a particular aircraft is of course unlimited, but aircraft specification sets the scene in allowing meaningful limits to be prescribed. A typical specification for a new or updated helicopter might contain the following requirements, exclusive of emergency operations such as personnel rescue and life saving:

- Prescribed missions, such as a hover role, a payload/range task or a patrol/loiter task. More than one are likely to be called for. A mission specification leads to a weight determination for payload plus fuel and thence to an all-up weight, in the standard fashion illustrated in Figure 7.1.
- Some specific atmosphere-related requirements, for example the ability to perform the mission at standard (ISA) temperature plus, say, 15°; the ability to perform a reduced mission at altitude; the ability to fly at a particular cruise speed.
- Specified safety requirements to allow for an engine failure.
- Specified environmental operating conditions, such as to and from ships or oil rigs.
- Prescribed dimensional constraints for stowage, air portability and so on.
- Possibly a prescribed power plant.

Calculations at the flexible design stage are only a beginning; as a design matures, more will be needed to check estimates against actual performance, find ways out of unexpected difficulties, or enhance achievement in line with fresh objectives. (One is effectively zeroing in on the details.)

Generally, in a calculation of achievable or required performance, the principal characteristics to be evaluated are:

1. Power needed in hover.
2. Power needed in forward flight.

Basic Helicopter Aerodynamics, Third Edition. John Seddon and Simon Newman.
© 2011 John Wiley & Sons, Ltd. Published 2011 by John Wiley & Sons, Ltd.

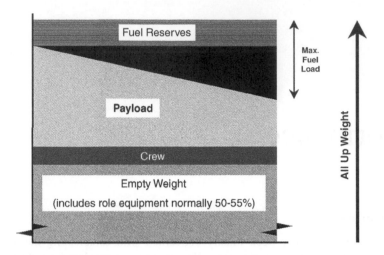

Figure 7.1 Determination of all-up weight for prescribed mission

3. Envelope of thrust limitations imposed by retreating-blade stall and advancing-blade compressibility drag rise.

The following sections concentrate on these aspects, using simple analytical formulae, mostly already derived. Item 3 must always be kept under review because the flight envelope so defined often lies inside the power limits and is thus the determining factor on level flight speed and manoeuvring capability.

A brief descriptive section is included on more accurate performance estimation using numerical methods. The chapter concludes with three numerical examples: the first concerns a practical achievement from advanced aerodynamics; the others are hypothetical, relating to directions in which advanced aerodynamics may lead in the future.

7.2 Hover and Vertical Flight

The formula relating thrust and power in vertical flight, according to blade element theory, was derived in Equation 3.54. The power required is the sum of induced power, related to blade lift, and profile power, related to blade drag. Converting to dimensional terms the equation is:

$$P = k_i(V_C + V_i)T + \frac{1}{8}C_{D0}\rho A_b V_T^3 \tag{7.1}$$

where in the induced term, using momentum theory as in Equation 2.28, one may write:

$$V_C + V_i = \frac{1}{2}V_C + \sqrt{\left(\frac{1}{2}V_C\right)^2 + \frac{T}{2\rho A}} \tag{7.2}$$

In the profile term, A_b is the total blade area, equal to sA, and V_T is the tip speed, equal to ΩR. This term is independent of the climb speed V_C; that is to say, the method gives the profile drag power as the same in climb as in hover.

If in Equation 7.1 the thrust is expressed in N, velocities in m/s, area in m^2 and air density in kg/m^3, the power is then in W or, when divided by 1000, in kW. Using imperial units, thrust in lb, velocities in ft/sec, area in ft^2 and density in slugs/ft^3 lead to a power in lb ft/sec or, on dividing by 550, to HP.

To make a performance assessment, Equation 7.1 is used to calculate separately the power requirements of main and tail rotors. For the latter, V_C disappears and the level of thrust needed is such as to balance the main rotor torque in hover; this requires an evaluation of hover trim, based on the simple equation:

$$T \cdot l = Q \tag{7.3}$$

where Q is the main rotor torque and l is the moment arm from the tail rotor shaft perpendicular to the main rotor shaft. The tail rotor power may be 10–15% of main rotor power. To these two are added allowances for transmission loss and auxiliary drives, perhaps a further 5%. This leads to a total power requirement, P_{req} say, at the main shaft, for a nominated level of main rotor thrust or vehicle weight. The power available, P_{av} say, is ascertained from engine data, debited for installation loss. Comparing the two powers determines the weight capability in hover, out of ground effect (OGE), under given ambient conditions. The corresponding capability in ground effect (IGE) can be deduced using a semi-empirical relationship such as Equation 2.64. The aircraft ceiling in vertical flight is obtained by matching P_{req} and P_{av} for nominated weights and atmospheric conditions.

In order to understand the sensitivity of power in climb to the climb rate, we recall Equation 2.27:

$$V_i^2 + V_C \cdot V_i - \frac{T}{2\rho A} = 0$$

From the solution, the combined induced (including k_i) and climb power becomes:

$$P_{i+C} = T\left[V_C - \frac{k_i V_c}{2} + \frac{k_i}{2}\sqrt{V_C^2 + \frac{2T}{\rho A}}\right] \tag{7.4}$$

Using V_0 as the hover induced velocity (for the given thrust) we have:

$$P_{i+C} = \frac{T}{2}\left[V_C(2-k_i) + k_i\sqrt{V_C^2 + 4V_0^2}\right] \tag{7.5}$$

from which we obtain, after partial differentiation with respect to V_C:

$$\frac{\partial P_{i+C}}{\partial V_C} = \frac{T}{2}\left[(2-k_i) + \frac{k_i V_C}{\sqrt{V_C^2 + 4V_0^2}}\right] \tag{7.6}$$

Hence for small finite quantities:

$$\frac{\Delta P}{\Delta V_C} = \frac{T}{2}\left[(2-k_i) + \frac{k_i \bar{V}_C}{\sqrt{\bar{V}_C^2 + 4}}\right] \tag{7.7}$$

where the overbar is used, as in Chapter 2, to denote normalization with respect to V_0 and the derivative is related to small increments in power and climb rate.

If this argument is examined in reverse, we have an equation which relates the climb rate achievable for a given excess of power.

Equation 7.7 now becomes:

$$\Delta V_{\mathrm{C}} = \frac{2\Delta P}{T\left[(2-k_{\mathrm{i}}) + \dfrac{k_{\mathrm{i}}\bar{V}_{\mathrm{C}}}{\sqrt{\bar{V}_{\mathrm{C}}^2 + 4}}\right]}$$

$$= \frac{\Delta P}{W}\cdot\left\{\left(\frac{2}{k_{\mathrm{D}}}\right)\cdot\frac{1}{\left[(2-k_{\mathrm{i}}) + \dfrac{k_{\mathrm{i}}\bar{V}_{\mathrm{C}}}{\sqrt{\bar{V}_{\mathrm{C}}^2 + 4}}\right]}\right\} \qquad (7.8)$$

$$= \frac{\Delta P}{W}\cdot\text{climb rate factor}$$

where the thrust is assumed to be equal to the aircraft weight in an equilibrium state. The k_{D} factor is applied to the thrust to take account of the fuselage download increased in the climb.

Taking k_{i} to be 1.15 and k_{D} to be 1.025, the term in braces in Equation 7.8 is plotted in Figure 7.2.

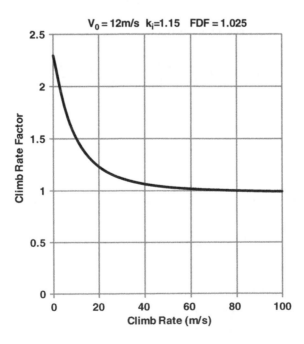

Figure 7.2   Climb rate factor

As the climb rate increases, the climb rate factor term drops from a value of approximately 2 in the hover to approximately 1 at a high climb rate. This is because, as the helicopter climbs away from hover, the climb causes a fall in the value of downwash, saving power which will contribute to that available from the engine(s). As the climb rate increases, the downwash will reduce to a very small value so this beneficial effect is lost and the engine power surplus is all that is available to generate a climb velocity. The two-to-one variation in factor between a zero climb rate and a high climb rate (say 6000 ft/min) is typical. Stepniewski and Keys (Vol. II, p. 55) suggest a linear variation between the two extremes. It should be borne in mind, however, that at low rates of either climb or descent, vertical movements of the tip vortices relative to the disc plane are liable to change the power relationships in ways which cannot be reflected by momentum theory and which are such that the power relative to that in hover is actually decreased initially in climb and increased initially in descent. These effects, which have been pointed out by Prouty [1], were mentioned in Chapter 2. Obviously in such situations Equation 7.4 and the deductions from it do not apply.

7.3   Forward Level Flight

The power–thrust relationship for level flight was derived in Chapter 5 and is given in idealized form in Equation 5.70, or with empirical constants incorporated in Equation 5.71. Generally we assume the latter form to be the more suitable for practical use and indeed to be adequate for most preliminary performance calculation. The equation shows the power coefficient to be the linear sum of separate terms representing, respectively, the induced power (rotor lift dependent), profile power (blade section drag dependent) and parasite power (fuselage drag dependent). It is in effect an energy equation, in which each term represents a separately identifiable sink of energy, and might have been calculated directly as such. In dimensional terms we have:

$$P = k_i V_i T + \frac{1}{8} C_{D0} \rho A_b V_T^3 \left[1 + k \left(\frac{V}{V_T} \right)^2 \right] + \frac{1}{2} \rho V^3 f \qquad (7.9)$$

in which V_T is the rotor tip speed, V the forward flight speed and f the fuselage-equivalent flat plate area, defined in Equation 5.88. The induced velocity V_i is given according to momentum theory by Equation 5.3; however, if we simplify the situation by assuming the rotor disc tilt is very small, Equation 5.3 can be rewritten as:

$$V_i = \frac{T}{2\rho A} \cdot \frac{1}{\sqrt{V^2 + V_i^2}} \qquad (7.10)$$

where V_Z is neglected as small order compared with V_i and V_X becomes V.
    The solution of (7.10) is given by:

$$V_i^2 = -\frac{1}{2} V^2 + \frac{1}{2} \sqrt{\left[V^4 + 4 \left(\frac{T}{2\rho A} \right)^2 \right]} \qquad (7.11)$$

Allowances should be added for tail rotor power and power to transmission and accessories: collecting these together in a miscellaneous item, the total is perhaps 15% of P at $V = 0$

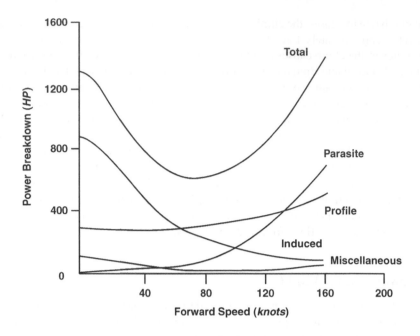

Figure 7.3 Typical power breakdown for forward level flight

(Section 7.2) and half this, say 8%, at high speed. Otherwise, if the evidence is available, the items may be assessed separately. The thrust T may be assumed equal to the aircraft weight W for all forward speeds above 5 m/s (10 knots) although at the highest speeds, the fuselage drag needs careful attention.

A typical breakdown of the total power as a function of flight speed is shown in Figure 7.3.

Induced power dominates the hover but makes only a small contribution in the upper half of the speed range. Profile power rises only slowly with speed unless and until the compressibility drag rise begins to be shown at high speed. Parasite power, zero in the hover, increases as V^3 and is the largest component at high speed, contributing about half the total. As a result of these variations the total power has a typical 'bucket' shape, high in the hover, falling to a minimum at moderate speed and rising rapidly at high speed to levels above the hover value. Except at high speed, therefore, the helicopter uses less power in forward flight than in hover.

Charts are a useful aid for rapid performance calculation. If power is expressed as P/δ, where δ is the relative air density at altitude, a power carpet can be constructed giving the variation of P/δ with W/δ and V. Figures 7.4a and b show an example, in which for convenience the carpet is presented in two parts, covering the low- and high-speed ranges. When weight, speed and density are known, the power required for level flight is read off directly.

The above is a relatively simple analysis of helicopter performance. The Appendix contains a fuller description of power prediction, the associated fuel consumption and mission analysis.

7.4 Climb in Forward Flight

As a first approximation let us assume that for climbing flight the profile power and parasite power remain the same as in level flight and only the induced power has to be reassessed.

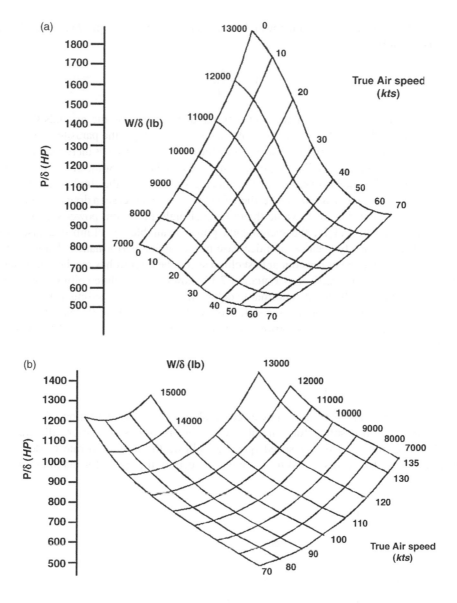

Figure 7.4 (a) Power carpet for rapid calculation (*low speed*). (b) Power carpet for rapid calculation (*high speed*)

The forced downflow alleviates the V_i term but the climb work term, TV_C, must be added. In coefficient form the full power equation is now:

$$C_P = k_i \lambda_i C_T + \frac{1}{8} C_{D0} s \left[1 + k\mu^2 \right] + \frac{1}{2} \rho \mu^3 \frac{f}{A} + \lambda_C C_T \qquad (7.12)$$

The usual condition for calculating climb performance is that of minimum power forward speed. In this flight regime, V_i is small compared with V, its variation from level flight to climb can be neglected and the incremental power ΔP required for climb is simply TV_C. Thus the rate of climb is given by:

$$V_C = \frac{\Delta P}{T} \qquad (7.13)$$

The result is a useful approximation but requires qualification on the grounds that since climbing increases the effective nose-down attitude of the fuselage, the parasite drag may be somewhat higher than in level flight. Also, because the main rotor torque is increased in climb – Equation 7.12 – an increase in tail rotor power is needed to balance it. Some of the incremental power available is absorbed in overcoming these increases and hence the climb rate potential is reduced, perhaps by as much as 30%. A further effect is that the increase in drag moves the best climb speed to a somewhat lower value than the level flight minimum power speed.

For a given aircraft weight, the incremental power available for climb decreases with increasing altitude, mainly because of a decrease in the engine performance characteristics. When the incremental power runs out at best climb speed the aircraft has reached its absolute ceiling at that weight. In practice, as Equation 7.13 shows, the absolute ceiling can only be approached asymptotically and it is normal to define instead a service ceiling as the height at which the rate of climb has dropped to a stated low value, usually about 0.5 m/s (100 ft/min). Increasing the weight increases the power required at all forward speeds and thereby lowers the ceiling.

7.4.1   Optimum Speeds

The bucket shape of the level flight power curve allows the ready definition of speeds for optimum efficiency and safety for a number of flight operations. These are illustrated in Figure 7.5. The minimum power speed (point A) allows the minimum rate of descent in autorotation. It is also, as discussed in the previous section, the speed for maximum rate of

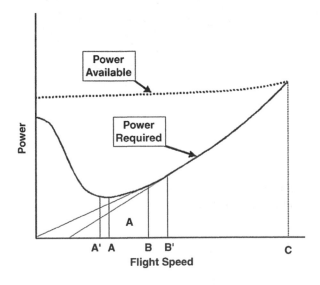

Figure 7.5   Optimum speeds and maximum speed

climb, subject to a correction to lower speed (A') if the parasite drag is increased appreciably by climbing. Subject to a further qualification, point A also defines the speed for maximum endurance or loiter time. Strictly the endurance relates directly to the rate of fuel usage, the curve of which, while closely similar to, is not an exact copy of the shaft power curve, owing to internal fuel consumption within the engine: the approximation is normally close enough to be acceptable.

Maximum glide distance in autorotative descent is achieved at speed B, defined by a tangent to the power curve from the origin. Here the ratio of power to speed is a minimum: the condition corresponds to that of gliding a fixed-wing aircraft at its maximum lift-to-drag ratio. Point B is also the speed for maximum range, subject to the fuel-flow qualification stated above. This is for the range in zero wind: in a headwind the best-range speed is at B', obtained by striking the tangent from a point on the speed axis corresponding to the wind strength. Obviously, for a tailwind the tangent is taken from a point on the negative speed axis, leading to a lower best-range speed than B. This is discussed in more detail in the Appendix.

7.5   Maximum Level Speed

The maximum speed attainable in level flight is likely to be limited by the envelope of retreating-blade stall and advancing-blade drag rise (Section 7.7). If and when power limited, it is defined by the intersection of the curves of shaft power required and shaft power available, (C) in Figure 7.5. In the diagram the power available has been assumed to be greater than that required for hover (out of ground effect) and, typically, to be nearly constant with speed, gaining a little at high speed from the effect of ram pressure in the engine intakes.

Approaching maximum speed, the power requirement curve is rising rapidly owing to the V^3 variation of parasite power. For a rough approximation one might suppose the sum of the other components, induced drag, profile drag and miscellaneous additional drag, to be constant and equal to, say, half the total. Then at maximum speed, writing P_{PARA} for the parasite power, we have:

$$P_{\mathrm{AV}} = 2P_{\mathrm{PARA}} = \rho V_{\mathrm{MAX}}^3 f \tag{7.14}$$

whence:

$$V_{\mathrm{MAX}} = \sqrt[3]{\frac{P_{\mathrm{AV}}}{\rho f}} \tag{7.15}$$

For a helicopter having 1000 kW available power, with a flat plate drag area of 1 m^2, the top speed at sea-level density would by this formula be 93.4 m/s (181 knots).

Increasing density altitude reduces the power available and may either increase or decrease the power required. Generally the reduction of available power dominates and V_{MAX} decreases. Increasing weight increases the power required (through the induced power P_{i}) without changing the power available, so again V_{MAX} is reduced.

7.6   Rotor Limits Envelope

The envelope of rotor thrust limits is the combination of operation on the blades at the stall boundary both at high angle of incidence and with the compressibility effects at high Mach number. Usually the restrictions occur within the limits of available power. The nature of the envelope is sketched in Figure 7.6.

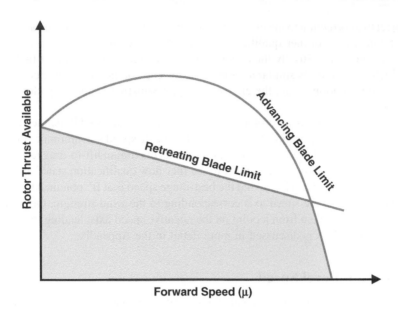

Figure 7.6   Nature of rotor thrust limits

In hover, conditions are uniform around the azimuth and blade stall sets a limit to the thrust available. As forward speed increases, maximum thrust on the retreating blade falls because of the drop in dynamic pressure (despite some increase in maximum lift coefficient with decreasing resultant Mach number) and this limits the thrust achievable throughout the forward speed range. By the converse effect, maximum thrust possible on the advancing side increases but is unrealizable because of the retreating-blade restriction. However, at higher speeds, as the advancing-tip Mach number approaches 1.0, its lift becomes restricted by shock-induced flow separation, leading to drag rise or pitching moment divergence, which eventually limits the maximum speed achievable. Thus the envelope comprises a limit on thrust from retreating-blade stall and a limit on forward speed from advancing-blade Mach effects. Without the advancing-blade problem, the retreating-blade stall would itself eventually set a maximum to the forward speed, as the figure-of-eight diagrams in Figure 6.1 show.

Calculation of the limits envelope is best done by computer, allowing the inclusion of sophisticated factors, natural choices among which are a non-uniform induced-velocity distribution, a compressibility factor on lift slope (usually $1/\beta$ where $\beta = (1-M^2)^{1/2}$, M being the blade section Mach number) and a representation of blade dynamic stall characteristics.

An example of the way in which the limits envelope can dominate performance issues is given later in Section 7.10.

7.7   Accurate Performance Prediction

The ability to deploy computer methods in performance calculation has been a major factor in the rapid development of helicopter technology since the Second World War. Results may often not be greatly different from those derived from the simple analytical formulae, but the fact that the feasibility of calculation is not dependent upon making a large number of challengeable assumptions is important in pinning down a design, making comparisons with flight tests or meeting guarantees. So it is that commercial organizations and research centres are equipped nowadays with computer programs for use in all the principal phases of performance calculation – hover characteristics, trim analysis, forward flight performance, rotor thrust limits and so on.

It may be noted *en passant* that performance calculation is generally not the primary factor determining the need for numerical methods. The stressing of rotor blades makes a greater demand for complexity in calculation. Another highly important factor is the need for quantification of handling characteristics, as for example to determine the behaviour of a helicopter flying in an adverse aerodynamic environment.

Within the realm of performance prediction are contained many sub-items, not individually dominant but requiring detailed assessment if maximum accuracy is to be achieved. One such sub-item is parasite drag, *in toto* an extensive subject, as with fixed-wing aircraft, about which not merely a whole chapter but a whole book could be written. For computation purposes the total drag needs to be broken down into manageable groupings, among which are streamlined and non-streamlined components, fuselage angle of attack, surface roughness, leakage and cooling-air loss. Maximum advantage must be taken of the review literature, as compiled by Hoerner [2], Keys and Wiesner [3] and others, and of background information on projects similar to the one in hand.

Once a best estimate of parasite drag has been made, the accuracy problem in power calculation devolves upon the induced and profile items, as Equation 7.12 shows, together with the additional sub-items of tail rotor power, transmission loss and power to auxiliaries. Improving the estimation of induced and profile power comes down to the ability to use a realistic distribution of induced velocity over the disc area and the most accurate blade section lift and drag characteristics, including dynamic effects. This information has to be provided separately; the problem in the rotor is then to ascertain the angles of attack and Mach numbers of all blade sections, these varying from root to tip and round the azimuth as the blade rotates. That is basically what the focal computer programs do. Iterative calculations are normally required among the basic equations of thrust, collective and cyclic pitch and the flapping angles. A primary difficulty with helicopter rotor analysis is that one cannot solve a subset of the rotor equations – one has to address the interactions all together. Starting with, say, values of thrust and the flapping coefficients, corresponding values of the pitch angle, collective and cyclic, can be calculated; the information then allows the blade angles and local Mach numbers to be determined, from which the lift forces can be integrated into overall thrust for comparison with the value initially assumed. When the iterations have converged, the required output data – power requirement, thrust limits, and so on – can be ascertained.

These brief notes provide an initial examination. Going more deeply into the subject would immerse one immediately into more detail, so a calculation strategy using spreadsheets is provided in the Appendix. In addition, an excellent and thorough exposition of the total process of performance prediction is available in Stepniewski and Keys (Vol. II), to which the reader who wishes to come to grips with the whole computational complex is also referred.

7.8 A World Speed Record

In the context of advanced rotor blade design as discussed in Chapter 6, and as an example of realized performance, it is of interest to record the capture of the world speed record for helicopters by a Westland Lynx aircraft in August 1986. The incentive to make the attempt was provided by the results of a programme of test flying on the Lynx fitted with an experimental set of blades in which lift-enhancing aerofoil sections of the RAE '96' Series (Section 6.2) were used throughout the length of the blade, together with the Westland tip design (Section 6.3) combining a sweepback benefit on local Mach number with delaying the tip stall. The tests showed the flight envelope to be improved by the equivalent of 35–40% increase in blade area and made it clear that level flight speeds beyond the existing record were achievable.

Different aerofoil sections were used for the inboard, mid-span and tip parts of the blade, chosen in relation to the local speed conditions and lift requirements. The section used for the tip was thinner than the other two. The blade was built in glass fibre with a single spar, special construction methods being employed.

The aircraft was a standard Lynx (Utility version) with a skid undercarriage, in which protuberance drag had been reduced to a minimum and an attempt had been made to reduce rotor head drag by fairings. The engine power was enhanced by water–methanol injection. The purpose of these measures was to ensure that, given a large alleviation in the flight envelope, the aircraft would not then be power limited unnecessarily.

For the record attempt, the course of 15 km was flown at 150 m above ground over the Somerset Levels, this being well within the altitude band officially required. The mean speed of two runs in opposite directions was 400.87 km/h (216 knots), exceeding the previous record by 33 km/h. The aircraft also had an extraordinarily good rate of climb near the bucket speed (80–100 knots), this being well over 20 m/s (4000 ft/min) – exceeding the capacity of the indicator instrument – and generally exhibited excellent flying characteristics. Figure 7.7 shows a photograph of the aircraft in flight and Figure 7.8 presents a spectacular view of the rotor blade.

Figure 7.7 World speed record helicopter in forward flight (Courtesy Agusta Westland)

Figure 7.8 BERP rotor blade on world speed record helicopter (G-LYNX) (Courtesy Agusta Westland)

7.9 Speculation on the Really Low-Drag Helicopter

The ideas in this section come mainly from M.V. Lowson [4]. It is of interest to consider, at least in a hypothetical manner, the lowest level of cruising power that might be envisaged for a really low-drag helicopter of the future, by comparison with levels typically achieved in current design. The demand for fuel-efficient operation is likely to increase with time, as more range-flying movements are undertaken, whether in an industrial or a passenger-carrying context. Any increase in fuel costs will narrow the operating cost differential between helicopters (currently dominated by maintenance costs) and fixed-wing aircraft, and the possibility of the helicopter achieving comparability is an intriguing one.

Reference to Figure 7.3 shows that at high forward speed, while all the power components need to be considered, the concept of a really low-drag (RLD) helicopter stands or falls on the possibility of a major reduction in parasite drag being achieved. This is not a priori an impossible task, since current helicopters have from four to six times the parasite drag of an aerodynamically clean fixed-wing aircraft. For the present exercise let us take as the data case a 4500 kg (10 000 lb) helicopter, the parasite drag of which, in terms of equivalent flat plate area, is broken down in Table 7.1. All calculations were made in imperial units and for simplicity

Table 7.1 Comparison of datum and target aircraft drag data

	Data aircraft (ft^2)	Target for RLD aircraft (ft^2)
Basic fuselage	2.74	2.3
Nacelles	0.80	0.4
Tail unit	0.45	0.3
Rotor head	4.29	0.8
Landing gear	1.55	0
Other	4.22	1.2
TOTAL	14.05	5.0

these are used in the presentation. The total, 14.05 ft^2, is somewhat higher than the best values currently achievable but is closely in line with the value of 19.1 ft^2 for an 18 000 lb helicopter used by Stepniewski and Keys (Vol. II) for their typical case.

In setting target values for the RLD helicopter, as given in Table 7.1, the arguments used are as follows. Minimum fuselage drag, inferred from standard texts such as Hoerner [2] and Goldstein [5], would be based on a frontal area drag coefficient of 0.05. This corresponds to 2 ft^2 flat plate area in our case, which is not strictly the lowest possible because helicopters traditionally have spacious cabins with higher frontal areas, weight for weight, than fixed-wing aircraft. A target value of 2.3 ft^2 is therefore entirely reasonable and might even be bettered. The reductions in nacelle and tail unit drags may be expected to come in time and with special effort. A large reduction in rotor head drag is targeted but the figure suggested corresponds to a frontal area drag coefficient about double that of a smooth ellipsoidal body, so while much work would be involved in reshaping and fairing the head, the target seems not impossible of attainment. Landing-gear drag is assumed to be eliminated by retraction or other means. In the miscellaneous item of the data helicopter, a substantial portion is engine-cooling loss, on which much research could be done. Tail rotor head drag can presumably be reduced in much the same proportion as that of the main rotor head. Roughness and protuberance losses will of course have to be minimized.

In total the improvement envisaged is a 64% reduction in parasite drag. Achievement of this target would leave the helicopter still somewhat inferior to an equivalent clean fixed-wing aircraft.

Such a major reduction in parasite drag will leave the profile power as the largest component of RLD power at cruise. The best prospect for reducing blade profile drag below current levels probably lies in following the lead given by fixed-wing technology in the development of supercritical aerofoil sections. Using such sections in the tip region postpones the compressibility drag rise to higher Mach number: thus a higher tip speed can be used which, by Equation 6.45, reduces the blade area required and thereby the profile drag. Advances have already been made in this direction, but whereas in the rotor design discussed in Chapter 6 a tip Mach number 0.88 was assumed, in fixed-wing research drag-rise Mach numbers as high as 0.95 were described by Haines [6]. Making up this kind of deficiency would reduce blade profile drag by about 15%. If it is supposed that in addition advances will be made in the use of thinner sections, a target of 20% lower profile power for the RLD helicopter seems reasonable.

Reduction in induced power will involve the use of rotors of larger diameter and lower disc loading than in current practice. Developments in blade materials and construction techniques will be needed for the higher aspect ratios involved. These can be expected, as can also the relaxation of some operational requirements framed in a military context, for example that of take-off in a high wind from a ship. A 10% reduction of induced power at cruise is therefore anticipated. The same proportion is assumed for the small residual power requirement of the miscellaneous items.

Table 7.2 shows the make-up of cruise power at 160 knots from Figure 7.3, representing the data aircraft, and compares this with the values for the RLD helicopter according to the foregoing analysis.

The overall reduction for the RLD helicopter is 41% of the power requirement of the data aircraft. An improvement of this magnitude would put the RLD helicopter into a competitive position with certain types of small, fixed-wing, propeller-driven business aircraft for low-altitude operation. Qualitatively it may be said that the RLD helicopter has a slightly higher parasite drag than the fixed-wing aircraft, about the same profile drag or slightly less (since the fixed-wing aircraft normally carries a greater wing area than is needed for cruise, while the

Table 7.2   Comparison of datum and target aircraft power data

	Data aircraft (HP)	RLD aircraft (HP)
Parasite	680	245
Profile	410	328
Induced	130	117
Misc.	80	72
TOTAL	1300	762

helicopter blade area can be made to suit, provided that adverse Mach effects are avoided) and a lower induced drag if the rotor diameter is greater than the fixed-wing span. The helicopter, however, has no ready answer to the ability of the fixed-wing aircraft to reduce drag by flying at high altitude. Equally of course, the fixed-wing aircraft cannot match the low-speed and hover capability of the helicopter.

7.10   An Exercise in High-Altitude Operation

Fixed-wing aircraft operate more economically at high altitude than at low. Aircraft drag is reduced and engine (gas-turbine) efficiency is improved, leading to increases in cruising speed and specific range (distance per unit of fuel consumed). With gas-turbine-powered helicopters, the incentive to realize similar improvements is strong: there are, however, basic differences to be taken into account. On a fixed-wing aircraft, the wing area is determined principally by the stalling condition at ground level; increasing the cruise altitude improves the match between area requirements at stall and cruise. On a helicopter, the blade area is fixed by a cruise speed requirement, while low-speed flight determines the installed power needed. The helicopter rotor is unable to sustain the specified cruising speed at altitudes above the density design altitude, the limitation being that of retreating blade stall. The calculations now to be described are of a purely hypothetical nature, intended to illustrate the kind of changes that could in principle convey a high-altitude flight potential. The altitude chosen for the exercise is 3000 m (10 000 ft), this being near the limit for zero pressurization. We are indebted to R.V. Smith for the work involved.

Imperial units are used as in the previous section. The data case is that of a typical light helicopter, of all-up weight 10 000 lb and having good clean aerodynamic design, though traditional in the sense of featuring neither especially low-drag nor advanced blade design. Power requirements are calculated by the simple methods outlined earlier in the present chapter. Engine fuel flow is related to power output in a manner typical of modern gas-turbine engines. Specific range (nautical miles per pound fuel) is calculated thus:

$$\text{Specific range(nm/lb)} = \text{forward speed((knots)/fuel)flow(lb/hr)}$$

A flight envelope of the kind described in Section 7.7 is assumed: this is primarily a retreating-blade limitation in which the value of W/δ (δ being the relative density at altitude) decreases from 14 000 lb at 80 knots to 8000 lb at 180 knots.

The results are presented graphically in Figures 7.9A–D.

Specific range is plotted as a function of flight speed for sea level (SL), 5000 ft and 10 000 ft altitude. Intersecting these curves are (a) the flight envelope limit, (b) the locus of best-range speeds and (c) the power limitation curve. We see that in case A, which is for the data helicopter, the flight envelope restricts the maximum specific range to 0.219 nm/lb, this

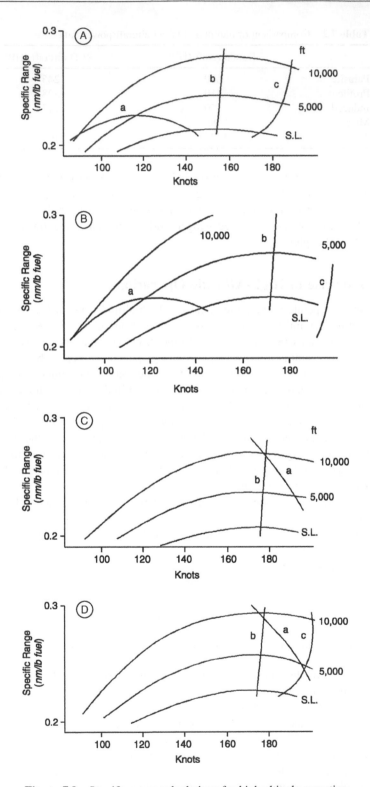

Figure 7.9 Specific range calculations for high-altitude operation

Table 7.3 Comparison of configuration performance

Data	Best range speed (knots)	Altitude (ft)	Specific range (nm/lb)	Weight penalty (lb)	Max. range (nm)	
					(1)	(2)
A	114	5000	0.219		357	357
B	120	4200	0.231	0		
C	174	10 000	0.267	652	274	458
D	174	10 000	0.293	225	433	503

occurring at 5000 ft and low speed (only 114 knots). So far as available power is concerned, it would be possible to realize the best-range speeds up to 10 000 ft and beyond.

Case B examines the effect of a substantial reduction in parasite drag. Using a less ambitious target than that envisaged in Section 7.9, a parasite drag two-thirds that of the data aircraft is assumed. At best-range speed a large increase in specific range at all altitudes is possible but, as before, the restriction imposed by the flight envelope is severe, allowing an increase to only 0.231 nm/lb, again at approximately 5000 ft and low speed (120 knots). It is clear that the full benefit of drag reduction cannot be realized without a considerable increase in rotor thrust capability. A comparison of cruising speeds emphasizes the deficiency: without the flight-envelope limitation the best-range speeds would be usable, that is at all heights a little above 150 knots for the data aircraft and 20 knots higher for the low-drag version.

The increase in thrust capability required by the low-drag aircraft to raise the flight envelope limit to the level of best-range speed at 10 000 ft is approximately 70%. Case C shows the performance of the low-drag aircraft supposing the increase to be obtained from the same percentage increase in blade area. Penalties of weight increase and profile power increase are allowed for, assumed to be in proportion to the area change. The best-range speed is now attainable up to over 9000 ft, while at 10 000 ft the specific range is virtually the same as at best-range speed, that is 0.267 nm/lb at 170 knots; this represents a 22% increase in specific range over the data aircraft, attained at 60 knots higher cruising speed.

For a final comparison, case D shows the effect of obtaining the required thrust increase by combination of a much smaller increase in blade area (24.5%) with conversion to an advanced rotor design, using an optimum distribution of cambered blade sections and the Westland advanced tip. The penalties in weight and profile power are thereby reduced considerably. The result is a further increase in specific range, to 0.293 nm/lb or 34% above that of the data aircraft, attained at the same cruising speed as in case C.

The changes are seen to further advantage by calculating also the maximum range achievable. This has been done in alternative ways, assuming that the weight penalty reduces (1) the fuel load or (2) the payload. On the first supposition, the weight penalty of case C results in a range reduction but with case D the gain more than compensates for the smaller weight penalty.

The characteristics of the various configurations are summarized in Table 7.3.

7.11 Shipborne Operation

The ability of a helicopter to take off and land vertically and to be able to hover efficiently makes it very suitable to operating from a deck on a ship. However, the location of the deck on the ship, its size and the fact that the ship will be at sea with its motion on the waves and the high winds it will experience make such operation very hazardous. However, the benefits of shipborne operation have made the development of the naval helicopter a subject of importance. To operate from the ship the helicopter must handle the following factors:

- The deck will be of limited size restricting the aircraft movements.
- The helicopter will have to perform its manoeuvres in high winds.
- The rotor downwash and the airflow over the ship will interact.
- The ship will be moving.
- The visibility of the pilot is very restricted both rearwards and downwards.

These factors affect the method of operation and the design of the aircraft. The touchdown of the landing on the deck is a major point of the landing. It is not subtle and the pilot will plant the helicopter down in a positive fashion. As with all naval aircraft, the vertical velocity at touchdown is usually of the order of double that of a landborne aircraft. This immediately places higher loads on the undercarriage and its mountings on the fuselage. The dynamic characteristics of the undercarriage legs must arrest the downward motion of the helicopter but also provide a very high level of damping as the undercarriage recoils as the axial loads diminish after touchdown. Taking these factors into account, the naval helicopter undercarriage is a sophisticated device which will carry a weight penalty. Having landed, the airframe is usually secured to the deck with a decklock arrangement. These are often carried on the aircraft itself, with the ship providing an attachment such as a steel wire net.

The approach and landing techniques vary, usually between countries. The following is that used by the Royal Navy. One aspect that needs to be emphasized is the lack of pilot's vision below and behind the aircraft. Additionally, with a two-seat layout in the cockpit, the pilot sits in the right-hand seat when viewed from behind. A typical landing and take-off are illustrated below (see Figure 7.10).

Figure 7.10   Landing trajectory for the Portuguese Lynx (Courtesy Agusta Westland)

Figure 7.11   A Lynx landing on deck showing the operation of the undercarriage (Courtesy Agusta Westland)

The approach is made on the port side of the ship with the aircraft coming in on a glide slope of about 3°. This brings the aircraft to a hover – relative to the ship – at the hangar height and 1.5 to 2 rotor radii off the hull and superstructure. A traverse is then made parallel to the hangar door until the helicopter is hovering above the ship's centreline. When the ship enters a quiescent state, the aircraft is lowered to the deck, with the aim of a firm touchdown (see Figure 7.10 and Figure 7.11). The undercarriage is designed to arrest the downward motion and also to kill off any tendency to recoil back upwards. If available, a reverse thrust can be selected for the main rotor, pressing the helicopter firmly onto the deck. The decklock is then activated providing a secure mechanical connection to the ship. The main rotor can then be returned to zero thrust. The undercarriage is often fitted with castoring wheels so the helicopter can be manoeuvred on deck using the tail rotor thrust to provide the turning moment.

One item of particular importance is the flow conditions surrounding the flight deck. Ship's flight decks are usually placed at the rear of the ship and with the hangar and hull sides, the flows are typically separating off sharp edges – that is bluff bodies. Figure 7.12 shows two flow types seen in a water tunnel experiment. Both flows are appropriate to wind perpendicular to the ship centreline. The upper image is just a hull and deck where the flow separates off the windward deck edge dividing the flow into a recirculating region above the deck surmounted by a clear airflow directed upwards. The lower image shows the effect of adding a hangar-type structure. There is now a separation line along the upwind hangar door edge. This combines with the original separation line to give a vortex flow which develops from the bottom door corner and covers the entire deck region.

If the flow is coming from the bow, the flow is influenced by the hangar as shown by Figure 7.13 which is a computational fluid dynamics (CFD) calculation.

The flow is dominated by the separation off the hangar roof and its eventual reattachment on the deck surface. In reality this reattachment point varies in position with respect to time. It is to this flow state that the helicopter enters as it traverses over the flight deck. The rotor downwash will interact with this flow producing a typical pattern as shown in Figure 7.14.

The flow is completely changed with the rotor downwash providing the main feature. There is now a significant recirculation between the front rotor edge and the hangar door. It is these

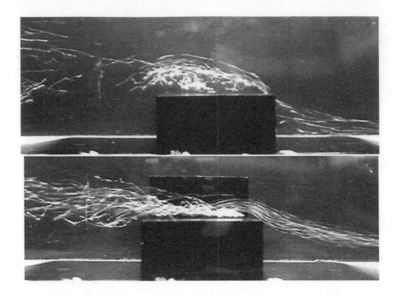

Figure 7.12   Flow past a hull and a hull–superstructure combination

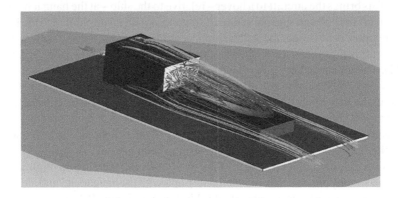

Figure 7.13   CFD predicted flow over bow – ship only

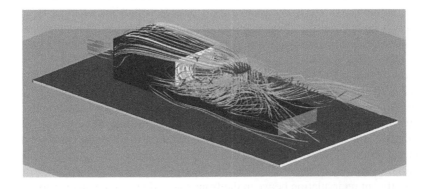

Figure 7.14   CFD predicted flow over bow – rotor and ship

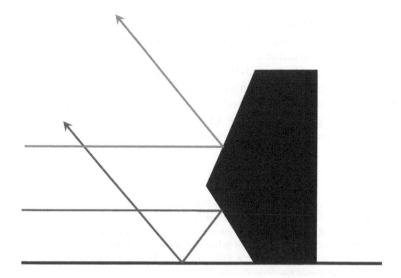

Figure 7.15   Stealth applied to ship profile

Figure 7.16   Profile of HMS *Dauntless*

effects which make shipborne operation so challenging for a helicopter. The separation off the ship's superstructure will be made more complex with the new breed of warship. In order to introduce stealth the superstructure is significantly changed.

Figure 7.15 shows two incident signals from a radar source and by using inclined surfaces the signals are returned in a specific direction which will minimize any returns to the enemy source. This alignment technique has been used on the more modern fighter aircraft designs. Figure 7.16 shows the flight deck of a Type 45 destroyer, HMS *Dauntless*, where the inclined faces of the hull and superstructure can be seen.

References

1. Prouty, R.W. (1985) *Helicopter Aerodynamics*, PJS Publications.
2. Hoerner, S.F. (1975) *Fluid Dynamic Drag*, published by author.
3. Keys, C.R. and Wiesner, R. (1975) Guidelines for reducing helicopter parasite drag. *JAHS*, **20**, 31.
4. Lowson, M.V. (1980) Thoughts on an efficient helicopter. *WHL Res. Memo.*
5. Goldstein, S. (1938) *Modern Developments in Fluid Dynamics*, Oxford University Press.
6. Haines, A.B. (1976) Aerodynamics. *Aero. J.*, July, 277–293.

8

Trim, Stability and Control[1]

8.1 Trim

The general principle of flight with any aircraft is that the aerodynamic, inertial and gravitational forces and moments about three mutually perpendicular axes are in balance at all times. In helicopter steady flight (non-rotating), the balance of forces determines the orientation of the main rotor in space. The balance of moments about the aircraft centre of gravity (CG) determines the attitude adopted by the airframe and when this balance is achieved, the helicopter is said to be trimmed. To a pilot the trim may be 'hands on' or 'hands off'; in the latter case, in addition to zero net forces and moments on the helicopter the control forces are also zero: these are a function of the internal control mechanism and will not concern us further, apart from a brief reference at the end of this section.

In deriving the performance equation for forward flight in Chapter 5 (Equation 5.70), the longitudinal trim equations were used in their simplest approximate form (Equations 5.66 and 5.67). They involve the assumption that the helicopter parasite drag is independent of fuselage attitude, or alternatively that Equation 5.70 is used with a particular value of D_P for a particular attitude, which is determined by solving a moment equation (see Figures 8.2a–c and the accompanying description below). This procedure is adequate for many performance calculations, which explains why the subject of trim was not introduced at that earlier stage. For the most accurate performance calculations, however, a trim analysis programme is needed in which the six equations of force and moment are solved simultaneously, or at least in longitudinal and lateral groups, by iterative procedures such as Stepniewski and Keys (Vol. II) have described.[2]

Consideration of helicopter moments has not been necessary up to this point in the book. To go further we need to define the functions of the horizontal tailplane and vertical fin and the nature of direct head moment.

---

[1] This chapter makes liberal use of unpublished papers by B. Pitkin, Flight Mechanics Specialist, Westland Helicopters.

[2] An illustration of the complexities introduced when a full 6 degree of freedom analysis is taken is the role of the tail rotor. It will be producing a thrust to balance the torque provided by the engine(s) to power the main rotor. This thrust will be a side force which will need to be reacted by a change in the lateral tilt of the main rotor disc. Hence the main and tail rotors have a mutual influence.

Basic Helicopter Aerodynamics, Third Edition. John Seddon and Simon Newman.
© 2011 John Wiley & Sons, Ltd. Published 2011 by John Wiley & Sons, Ltd.

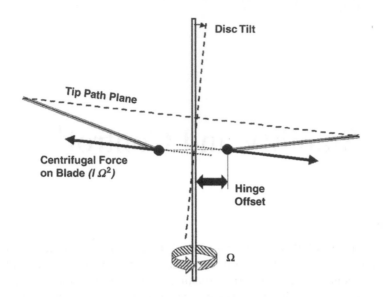

Figure 8.1   Direct rotor moment

In steady cruise the function of a tailplane is to provide a pitching moment to offset that produced by the fuselage and thereby reduce the net balancing moment which has to be generated by the rotor. The smaller this balancing moment can be, the less is the potential fatigue damage on the rotor. In transient conditions the tailplane pitching moment is stabilizing, as on a fixed-wing aircraft, and offsets the inherent static instability of the fuselage and to some extent that of the main rotor. A fixed tailplane setting is often used, although this is only optimum (fuselage attitude) for one combination of flight condition and CG location.

A central vertical fin is multi-functional: it generates a stabilizing yawing moment and also provides a structural mounting for the tail rotor. The central fin operates in a poor aerodynamic environment, as a consequence of turbulent wakes from the main and tail rotors and blanking by the fuselage, but fin effectiveness can be improved by providing additional fin area near the tips of the horizontal tailplane.[3]

When the flapping hinge axis is offset from the shaft axis (the normal condition for a rotor with three or more blades), the centrifugal force on a blade produces (Figure 8.1) a pitching or rolling moment proportional to disc tilt. Known as direct rotor moment, the effect is large because, although the moment arm is small, the centrifugal force is large compared with the aerodynamic and inertial forces.[4] A hingeless rotor produces a direct moment perhaps four times that of an articulated rotor for the same disc tilt. Analytically this would be expressed by, according to the flexible element, an effective offset four times the typical 3–4% span offset of the articulated hinge.

Looking now at a number of trim situations, in hover with zero wind speed the rotor thrust is vertical in the longitudinal plane, with magnitude equal to the helicopter weight corrected for

---

[3] This is seen in extreme form with the Westland Lynx world speed record helicopter (G-LYNX), see Figure 7.7.
[4] Main rotor blade tip accelerations of 750–1000g are typical.

fuselage downwash. For accelerating away from hover the rotor disc must be inclined forward and the thrust magnitude adjusted so that it is equal to, and directly opposed to, the vector sum of the weight and the inertial force due to acceleration. In steady forward flight the disc is inclined forward and the thrust magnitude is adjusted so that it is equal to, and directly opposed to, the vector sum of the weight and aerodynamic drag.

The pitch attitude adopted by the airframe in a given flight condition depends upon a balance of pitching moments about the CG. Illustrating firstly without direct rotor moment or tailplane and airframe moment, the vector sum of aircraft drag (acting through the CG) and weight must lie in the same straight line as the rotor force. This direction being fixed in space, the attitude of the fuselage depends entirely upon the CG position. With reference to Figures 8.2a and b, a forward location results in a more nose-down attitude than an aft location. The effect of a direct rotor moment is illustrated in Figure 8.2c for a forward CG location. Now the rotor thrust and resultant force of drag and weight, again equal in magnitude, are not in direct line but must be parallel, creating a couple which balances the other moments. A similar situation exists in the case of a net moment from the tailplane and airframe. For a given forward CG position, the direct moment makes the fuselage attitude less nose-down than it would otherwise be. Reverse results apply for an aft CG position. At high forward speeds, achieving a balanced state may involve excessive nose-down attitudes unless the tailplane can be made to supply a sufficient restoring moment.

Turning to the balance of lateral forces, in hover the main rotor thrust vector must be inclined slightly sideways to produce a force component balancing the tail rotor thrust. This results in a hovering attitude tilted 2° or 3° to port (Figure 8.3). In sideways flight the tilt is modified to balance sideways drag on the helicopter: the same applies to hovering in a crosswind. In forward flight the option exists, by sideslipping to starboard, to generate a sideforce on

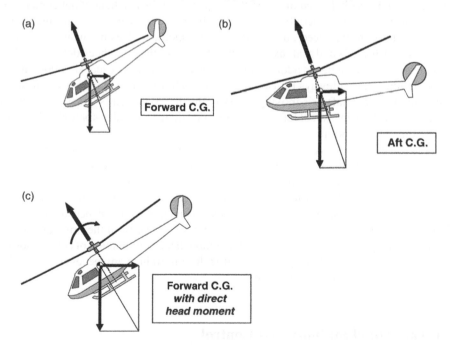

Figure 8.2 Fuselage attitude in forward flight: (a) forward CG; (b) aft CG; (c) forward CG with direct head moment

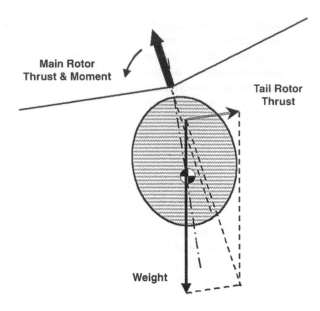

Figure 8.3 Lateral tilt in hover

the airframe which, at speeds above about 50 knots, will balance the tail rotor thrust and allow a zero-roll attitude to be held.

With the lateral forces balanced in hover, the projection of the resultant of helicopter weight and tail rotor thrust will not generally pass through the main rotor centre, so a rolling couple is exerted which has to be balanced out by a direct rotor moment. This moment depends upon the angle between disc axis and shaft axis and since the first of these has been determined by the force balance, the airframe has to adopt a roll attitude to suit. For the usual situation, in which the line of action of the sideways thrust component is above that of the tail rotor thrust, the correction involves the shaft axis moving closer to the disc axis; that is to say, the helicopter hovers with the fuselage in a small left-roll attitude. Positioning the tail rotor high (close to hub height) minimizes the amount of left-roll angle needed.

Yawing moment balance is provided at all times by selection of the tail rotor thrust, which balances the combined effects of main rotor torque reaction, airframe aerodynamic yawing moment due to sideslip and inertial moments present in manoeuvring.

The achievement of balanced forces and moments for a given flight condition is closely linked with stability. An unstable aircraft theoretically cannot be trimmed, because the slightest disturbance, atmospheric or mechanical, will cause it to diverge from the original condition. A stable aircraft may be difficult to trim because, although the combination of control positions for trim exists, over-sensitivity may make it difficult to introduce any necessary fine adjustments to the aerodynamic control surfaces.

8.2 Treatment of Stability and Control

As with a fixed-wing aircraft, both static stability and dynamic stability contribute to the flying qualities of a helicopter. Static stability refers to the initial tendency of the aircraft to return to its trimmed condition following a displacement. Dynamic stability considers the subsequent motion in time, which may consist of a dead-beat return, an oscillatory return, a no-change motion, an oscillatory divergence or a non-return divergence; the first two signify positive stability, the third neutral stability and the last two negative stability (instability). A statically unstable motion is also dynamically unstable but a statically stable motion may be either stable or unstable dynamically.

The subject of stability and control in totality is a formidable one. The part played by the rotor is highly complicated, because strictly each blade possesses its own degrees of freedom and makes an individual contribution to any disturbed motion. Fortunately, however, analysis can almost always be made satisfactorily by considering the behaviour of the rotor as a whole. Even so it is useful to make additional simplifying assumptions: those which pave the way for a classical analysis, similar to that made for fixed-wing aircraft, come essentially from the work of Hohenemser [1] and Sissingh [2] and are the following:

- in disturbed flight the accelerations are small enough not to affect the rotor response, in other words the rotor reacts in effect instantaneously to speed and angular rate changes;
- rotor speed remains constant, governed by the engine;
- longitudinal and lateral motions are uncoupled so can be treated independently. (*Strictly speaking, longitudinal and lateral motions are in fact coupled. However, they can be considered uncoupled for a first analysis. Examples of situations where coupling is significant are:*
  - *roll manoeuvres;*
  - *lateral disc tilt induced by forward flight;*
  - *tail rotor thrust.*

Cross-coupling is present in situations such as these and has significant effects on the handling qualities.)

Given these important simplifications, the mathematics of helicopter stability and control is nevertheless heavy (Bramwell's Chapter 7), edifying academically but hardly so otherwise, and in practice strongly dependent upon the computer for results. In this chapter we shall be content with descriptive accounts, which bring out the physical characteristics of the motions involved.

No absolute measure of stability, static or dynamic, can be stipulated for helicopters in general, because flying qualities depend on the particular blend of natural stability, control and autostabilization. Also, stability must be assessed in relation to the type of mission to be performed.

8.3   Static Stability

We consider the nature of the initial reaction to various forms of disturbance from equilibrium. Longitudinal and lateral motions are treated independently. The contributions of the rotor to forces and moments arise from two sources, variations in magnitude of the rotor force vector and variations in the inclination of this vector associated with disc tilt, which is defined by the blade flapping motion. This motion is highly dependent on hinge offset and blade Lock number – effectively the rotor control power. This is the ability to generate a moment about the rotor hub by the application of cyclic pitch and hence induce blade flapping which defines the rotor disc tilt.

8.3.1   Incidence^5 Disturbance

An upward imposed velocity (e.g. a gust) increases the incidence of all blades, giving an overall increase in thrust magnitude. Away from hover, the dissimilarity in relative airspeed on the advancing and retreating sides leads to an incremental flapping motion, which results in a nose-up tilt of the disc. Since the rotor centre lies above the aircraft CG, the pitching moment caused by the change of inclination is in a nose-up sense, that is destabilizing and increasingly so with increase of forward speed. In addition, the change in thrust magnitude itself generates a moment contribution, the effect of which depends upon the fore and aft location of the CG relative to the rotor centre. In a practical case, the thrust vector normally passes ahead of an aft CG location and behind a forward one, so the increase in thrust magnitude aggravates the destabilizing moment for an aft CG position and alleviates it for a forward one. The important characteristic therefore is a degradation of longitudinal static stability with respect to incidence, at high forward speed in combination with an aft CG position. This is also reflected in a degradation of dynamic stability under the same flight conditions. It should be noted that these fundamental arguments relate to rigid blades. With the advent of modern composite materials for blade construction, judicious exploitation of the distribution of inertial, elastic and aerodynamic loadings allows the possibility of tailoring the blade aeroelastic characteristics to alleviate the inherently destabilizing features just described.

Of the other factors contributing to static stability, the fuselage is normally destabilizing in incidence, a characteristic of all streamlined three-dimensional bodies. Hinge offset, imparting an effective stiffness, likewise aggravates the incidence instability. The one stabilizing contribution comes from the horizontal tailplane. Figure 8.4 represents the total situation

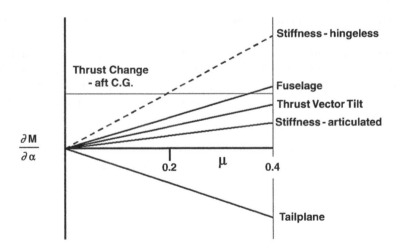

Figure 8.4   Contributions to static stability incidence

---

[5] The term 'incidence' requires a brief reflection. In fixed-wing terminology, it is often defined as the orientation of a data line in the wing/blade relative to the forward flight direction. Helicopter aerodynamics tends to use the term 'pitch angle' for this orientation and reserves incidence for the inclination of the data line to the local incident flow including any downwash velocity. This is consistent with the term 'angle of attack'. For the present discussion, incidence is the angle of the fuselage to the airflow direction generated by the motion of the aircraft.

diagrammatically. The tailplane compensates for the inherent instability of the fuselage, leaving the rotor contributions as the determining factors. Of these, the stiffness effect for an articulated rotor is generally of similar magnitude to the thrust vector tilt moment. With a hingeless rotor (Section 8.5) the stiffness effect is much greater. The effect is denoted in this figure as M_α; however, this effect is normally associated with a vertical velocity perturbation and so the derivative M_W is frequently used in this context.

8.3.2   Forward Speed Disturbance

An increase in forward speed leads to incremental flapping, resulting in a change in nose-up disc tilt. The amount of change is reckoned to be about $1°$ per $10\,m/s$ speed increase, independently of the flight speed. The thrust vector is effectively inclined rearwards, supported by the nose-up pitching moment produced, providing a retarding force component and therefore static stability with respect to forward speed. This characteristic is present in the hover but nevertheless contributes to a dynamic instability there (see Section 8.4.2).

An increase in speed causes the airframe drag to rise and this contributes, more effectively with initial forward speed, to a positive speed–stability characteristic for the helicopter, except in the hover.

8.3.3   Angular Velocity (Pitch or Roll Rate) Disturbance

The effect of a disturbance in angular velocity (pitch or roll) is complex. In brief, a gyroscopic moment about the flapping hinge produces a phased flapping response and the disc tilt resulting from this generates a moment opposing the particular angular motion. Thus the rotor exhibits damping in both pitch and roll. Moments arising from non-uniform incidence over the disc lead to cross-coupling, that is rolling moment due to rate of pitch and vice versa.

8.3.4   Sideslip Disturbance

In a sideslip disturbance, the rotor 'sees' a wind unchanged in velocity but coming from a different direction. As a result the direction of maximum flapping is rotated through the angle of sideslip change and this causes a sideways tilt of the rotor away from the wind. There is therefore a rolling moment opposing the sideslip, corresponding effectively to the dihedral action of a fixed-wing aircraft. In addition the sideslip produces a change in incidence of the tail rotor blades, so that the tail rotor acts like a vertical fin providing 'weathercock' stability.

8.3.5   Yawing Disturbance

A disturbance in yaw causes a change of incidence at the tail rotor and so again produces a fin damping effect, additional to that of the actual aircraft fin. Overall, however, basic directional stability tends to be poor because of degradation by upstream flow separations and wake effects.

8.3.6   General Conclusion

It is seen from the above descriptions that longitudinal static stability characteristics are significantly different from, and more complex than, those of a fixed-wing aircraft, while lateral characteristics of the two types of aircraft are similar, although the forces and moments arise in different ways.

8.4   Dynamic Stability

8.4.1   Analytical Process

The mathematical treatment of dynamic stability given by Bramwell follows the lines of the standard treatment for fixed-wing aircraft. Wind axes are used, with the X axis parallel to the flight path, and the stability derivatives ultimately are fully non-dimensionalized. The classical format is useful because it is basic in character and displays essential comparisons prominently. The most notable distinction which emerges is that, whereas with a fixed-wing aircraft the stability quartic equation splits into two quadratics, leading to a simple physical interpretation of the motion, with the helicopter this unfortunately is not so and as a consequence the calculation of roots becomes a more complicated process.

Industrial procedures for the helicopter tend to be on rather different lines. The analysis is generally made with reference to body axes, with origin at the CG. In this way the X axis remains forward relative to the airframe, whatever the direction of flight or of relative airflow. The classical linearization of small perturbations is still applicable in principle, the necessary inclusion of initial-condition velocity components along the body axes representing only a minor complication. Force and moment contributions from the main rotor, tail rotor, airframe and fixed tail surface are collected along each body axis, as functions of flow parameters, control angles and flapping coefficients, and are then differentiated with respect to each independent variable in turn. In earlier days, computational techniques provided ready solutions to the polynomials. However, computer hardware and software have improved to an extent where many different techniques for solving the equations are possible. Full non-dimensionalization of the derivatives is less useful than for fixed-wing aircraft and a preferred alternative is to 'normalize' the force and moment derivatives in terms of the helicopter weight and moment of inertia respectively. This means that linear and rotational accelerations become the yardstick. These normalized terms are often referred to as *concise derivatives*.

8.4.2   Special Case of Hover

In hovering flight the uncoupled longitudinal and lateral motions break down further. Longitudinal motion resolves into an uncoupled vertical velocity mode and an oscillatory mode coupling forward velocity and pitch attitude. In a similar manner, lateral motion breaks down into an uncoupled yaw mode and an oscillatory mode coupling lateral velocity and roll attitude. Both of these coupled modes are dynamically unstable. The physical nature of the longitudinal oscillation is illustrated in Figure 8.5 and can be described as follows.

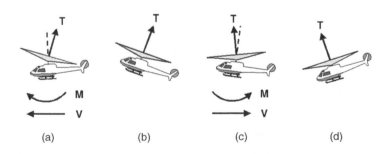

Figure 8.5   Longitudinal dynamic instability in hover

Suppose the hovering helicopter was to experience a small forward velocity as at (a). This would usually be the effect of a small horizontal gust impinging on the aircraft. Incremental flapping creates a nose-up disc tilt, which results in a nose-up pitching moment on the aircraft. This is as described in Section 8.3.2 (the important overall qualification being that there is no significant aircraft drag force). A nose-up attitude develops and the backward-inclined thrust opposes the forward motion and eventually arrests it, as at (b). The disc tilt relative to the rotor shaft and hence the rotor moment have now been reduced to zero. A backward swing commences, in which the disc tilts forward, exerting a nose-down moment, as at (c). A nose-down attitude develops and the backward movement is ultimately arrested, as at (d). The helicopter then accelerates forward under the influence of the forward inclination of thrust and returns to the situation at (a). Mathematical analysis shows, and experience confirms, that the motion is dynamically unstable, the amplitude increasing steadily if the aircraft is left to itself.

This longitudinal divergent mode and its lateral–directional counterpart constitute a fundamental problem of hovering dynamics. They require constant attention by the pilot, though since both are usually of low frequency, some degree of instability can generally be allowed. It remains the situation, however, that 'hands-off' hovering is not possible unless a helicopter is provided with an appropriate degree of artificial stability.

8.5   Hingeless Rotor

A hingeless rotor flaps in similar manner to an articulated rotor and both the rotor forces and the flapping derivatives are little different between the two. Terms expressing hub moments, however, are increased severalfold with the hingeless rotor so that, as has been said, compared with the 3% to 4% hinge offset of an articulated rotor, the effective offset of a hingeless rotor is likely to be 12% to 16% or even higher. This increased stiffness has an adverse effect on longitudinal static stability: in particular the pitch instability at high speed is much more severe (Figure 8.4). A forward CG position is an alleviating factor, but in practice the CG position is dominated by role considerations. The horizontal tailplane can be designed to play a significant part. Not only is the stabilizing influence a direct function of tailplane size, but also the angular setting to the fuselage affects the pitching moment balance in trim and can be used to minimize hub moment over the critical part of the operational flight envelope. Despite this, however, the stability degradation in high-speed flight normally remains a dominant feature.

8.6   Control

Control characteristics refer to a helicopter's ability to respond to control inputs and so move from one flight condition to another. The inputs are made, as has been seen, by applying pitch angles to the rotor blades so as to generate the appropriate forces and moments. On the main rotor the angles are made up of the collective pitch θ_0 and the longitudinal and lateral cyclic pitch angles B_1 and A_1 as introduced in Chapter 4. The tail rotor conventionally has only collective pitch variation, determined by the thrust required for yawing moment balance.

As already introduced in Section 8.3, when the helicopter experiences a rate of pitch, the rotor blades are subjected to gyroscopic forces proportional to that rate. A nose-up rotation induces a download on an advancing blade, leading to nose-down tilt of the rotor disc. The associated offset of the thrust vector from the aircraft CG and the direct rotor moment are both in the sense opposing the helicopter rotation and constitute a damping effect or stabilizing feature. A similar argument applies to the gyroscopic effects of a rate of roll.

Adequacy of control is formally assessed in two ways, by control power and control sensitivity. Control power refers to the moment that can be generated for a given control input. It is effectively the slope of the moment v control input curve. Normalizing this in terms of aircraft moment of inertia, the measure becomes one of initial acceleration produced per unit displacement of the cyclic control stick. Control sensitivity recognizes the importance of a correlation between control power and the damping of the resultant motion – it reflects the maximum slope of the timewise response to the control input – and the ratio can be expressed as angular velocity per unit stick displacement. High control sensitivity means that control power is large relative to damping, so that a large angular velocity is reached before the damping moment stabilizes the motion.

The large effective offset of a hingeless rotor conveys both increased control power and greater inherent damping, resulting in shorter time constants and crisper response to control inputs. Basic flying characteristics in the hover and at low forward speeds are normally improved by this, because the more immediate response is valuable to the pilot for overcoming the unstable oscillatory behaviour described in Section 8.4.2. The ability of the hingeless rotor to manoeuvre the helicopter with a higher control power indicates a better path for transmitting moments and forces. This, unfortunately, causes the problem of better transmission of vibration.

A mathematical treatment of helicopter response is given by Bramwell (pp. 231–249) and illustrated by typical results for a number of different control inputs. His results for the normal acceleration produced by a sudden increase of longitudinal cyclic pitch (B_1) in forward level flight at advance ratio 0.3 are reproduced in Figure 8.6. We note the more rapid response of the hingeless rotor compared with the articulated rotor, a response which the equations show to be divergent in the absence of a tailplane. Fitting a tailplane reduces the response rates and in both cases appears to stabilize them after 3 or 4 seconds.

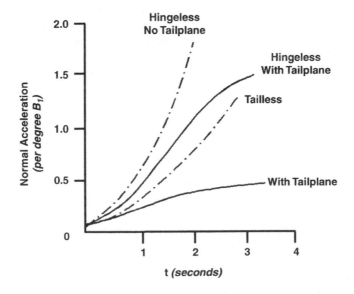

Figure 8.6   Calculated rotor response to B_1 (after Bramwell)

Roll response in hover is another important flying quality, particularly in relation to manoeuvring near the ground. In an appropriate example, Bramwell shows the hingeless helicopter reaching a constant rate of roll within less than a second, while the articulated version takes 3 or 4 seconds to do so. For a given degree of cyclic pitch, the final roll rates are the same, because the control power and roll damping differ in roughly the same proportion in the two aircraft.

Rotor response characteristics can be described more or less uniquely in terms of a single non-dimensional parameter, the stiffness number S, defined as:

$$S = \frac{\left(\lambda_\beta^2 - 1\right)}{n} \qquad (8.1)$$

This expresses the ratio of elastic to aerodynamic flapping moments on the blade. λ_β is the blade natural flapping frequency, having the value 1.0 for zero blade offset and related generally to the percentage offset e by:

$$\lambda_\beta^2 = 1 + \frac{3e}{2} \qquad (8.2)$$

Thus a 4% offset yields a value $\lambda_\beta = 1.03$; for hingeless rotors the λ_β values are generally in the range 1.09 to 1.15. In Equation 8.1, n is a normalizing inertia number. Some basic rotor characteristics are shown as functions of stiffness number in Figure 8.7.

Taking the four parts of the diagram in turn, the following comments can be made.

(a) Rotors have until now made use of only relatively restricted parts of the inertia/stiffness plane.
(b) In the amount of disc tilt produced on a fixed hovering rotor per degree of cyclic pitch, articulated and the 'softer' hingeless rotors are practically identical.
(c) On the phase lag between cyclic pitch application and blade flapping, we observe the standard 90° for an articulated rotor with zero hinge offset (the teetering rotor), decreasing with increase of offset, real or effective, to 15°–20° lower for a hingeless rotor.
(d) For the low-stiffness numbers of articulated rotors, the principal component of moment about the aircraft CG is likely to be that produced by thrust vector tilt. Hingeless rotors, however, produce moments mainly by stiffness; their high hub moment gives good control for manoeuvring but needs to be minimized for steady flight, in order to restrict as much as possible hub load fluctuations and vibratory input to the helicopter.

8.7 Autostabilization

In order to make the helicopter a viable operational aircraft, shortcomings in stability and control characteristics generally have to be made good by use of automatic flight control systems. The complexity of such systems, providing stability augmentation, long-term datum-holding autopilot functions, automatically executed manoeuvres and so on, depends upon the mission task, the failure survivability requirements and of course the characteristics of the basic helicopter.

Autostabilization is the response to what is perhaps the commonest situation, that in which inadequate basic stability is combined with ample control power. The helicopter is basically flyable but in the absence of automatic aids, continuous correction by the pilot would be

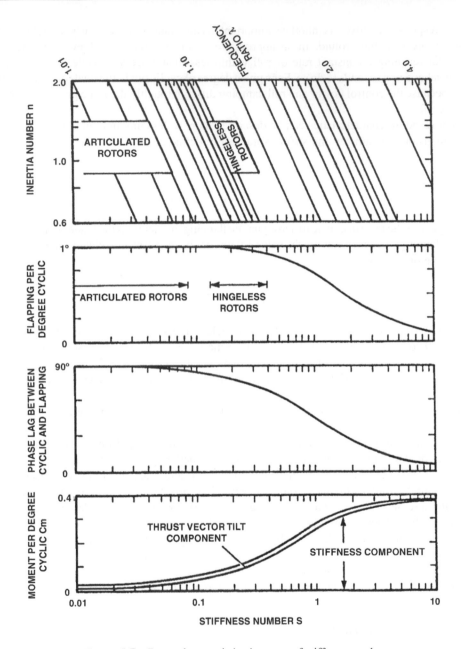

Figure 8.7 Rotor characteristics in terms of stiffness number

required – a tiring process and in some conditions (such as flying on instruments) potentially dangerous. The corrective is to utilize some of the available control power to generate moments proportional to a given motion variable and thereby correct the motion. An automatic signal is superimposed on the pilot's manual input, without directly affecting it. No signal feeds back to the controls; the pilot merely experiences the changed flying character.

Autostabilizing systems have in the past used mechanical devices integral to the rotor; typical of these are the Bell stabilizer bar and the Lockheed control gyro (see Figure 1.30).

Alternatively, devices may be electromechanical, operating on attitude or rate signals from helicopter motion sensors. Electrical or electronic systems are the more flexible and multi-purpose. An example is the attitude hold system, which returns the helicopter always to the attitude commanded, even in disturbing environments such as gusty air. Naturally, the more the stability is augmented in this way, the greater the attention that has to be paid to augmenting the control power remaining to the pilot. The balance is often achieved by giving the pilot direct control over the attitude datum commanded. The design of a particular system is governed by the degree of augmentation desired and the total control power available.

The subject of helicopter stability and handling qualities is a very involved one and this chapter can only provide an entry point for the reader. For further information the reader is encouraged to consult dedicated texts on this subject such as Padfield as listed in Chapter 1.

References

1. Hohenemser, K. (1939) Dynamic stability of a helicopter with hinged rotor blades, NACA Tech. Memo. 907.
2. Sissingh, G.J. (1946) Contributions to the dynamic stability of rotary wing aircraft with articulated blades, Air Materiel Command Trans. F-TS-690-RE.

9

A Personal Look at the Future

Somebody once referred to the future thus:

There is no future – it is just the past repeating itself!

In some ways this could be applied to the evolution of rotary-wing aircraft. The tilt wing, tilt rotor, compound and stopped rotor configurations have all come and gone and come again. In some particular instances a configuration is adopted by a manufacturer who has pursued it with a continuous programme of research and development. There will be many reasons why these things happen, but the learning process must play a part. Technologies, such as electronics and computing power, allied to a constantly improving set of methods open doors previously closed to the helicopter designer/engineer. These opening doors will encourage the revisiting of the various configurations which in most cases are devoted to ways of getting past the aerodynamic speed limit of the conventional main rotor system.

Any comments on the future have to express opinions particular to the author. I do not intend to cause upset and, particularly, do not wish to tread on any sensitive toes. If, however, I provoke argument – then that is healthy. You may think me wrong, but I hope you acknowledge that my comments are expressed with the best will in the world.

What follows is the result of consulting several papers [1–3] written on areas defining where rotary-wing development can be considered.

The experience of the 1982 Falkland Islands conflict showed a level of rotary-wing operations far exceeding the operations of any other type of aircraft.

The introduction of gas-turbine propulsion systems around 1960 began the move away from piston engines. The success of these prime-movers can be attributed to several factors, namely:

- Fuel efficiency.
- Advantages in use of internal space.
- The torque transferred to the transmission was of a smoother character with fewer fluctuations.
- No clutch was required.

Basic Helicopter Aerodynamics, Third Edition. John Seddon and Simon Newman.
© 2011 John Wiley & Sons, Ltd. Published 2011 by John Wiley & Sons, Ltd.

These were in spite of considerably higher speed reductions.

Then, the introduction of a rotor speed governing system, replacing the hand throttle usually placed on the collective lever control. The rotor is best served by rotating at a constant rotor speed. This is for reasons of performance but, equally importantly, to permit vibration suppression systems to have a fixed set of rotor vibration frequencies which allows vibration control to be more effective.

The use of electric power is under close scrutiny, perhaps initially for the tail rotor. The transmission can also be carefully reassessed. Gear contact is a source of wear and potential difficulties so recently research is being directed at magnetic gear contact making way for a completely new transmission layout. The skeletal gearbox was examined in the 1970s when the load bearing of the main rotor gearbox was taken from the enclosing structure to a direct line through the gear shafts themselves. There was then a rudimentary enclosure – not required to withstand the helicopter weight – which effectively kept the rain out and the lubricant in.

Automatic flight control systems (AFCSs) appeared towards the end of the 1950s. Their immediate benefit is the release of the pilot from controlling the engine throttle, thereby providing a reduction in their workload. This permits the pilot to focus on the various tasks they have to accomplish during the particular operational sortie. The initial control was provided by analogue systems known as automatic stabilization equipment (ASE) which had a relatively modest 10% authority. To this were added height and speed locks for use in cruise.

In more recent times the AFCS is now based on digital processing – albeit with a similarly limited control authority. To move on further, the control of the aircraft by such a system must earn the trust and confidence of the rotorcraft industry and operators and only then can the control system be used to mould the handling qualities of new aircraft together with new operations.

Improvements in assessing and modelling the helicopter rotor are a constant source of effort. The helicopter rotor is a very demanding place and any experiments conducted on this require a robust and accurate instrumentation system. This will need a considerable amount of computing power to process the streaming data. This computing power is also providing opportunities to investigate the aerodynamic and dynamic features of the rotor. For instance, the interaction of blades with the vortex wake can be more closely examined and the blade structural dynamic behaviour, governed by the natural modes of vibration, can be determined with more confidence and increasing detail. Modern aerodynamic modelling requires the most subtle of blade characteristics to be known. It behoves such work to establish the blade dynamic behaviour when subjected to rapidly changing aerodynamic forces if further progress is to be made. However, one must always be prepared for the unexpected, and unacceptable vibration characteristics of the aircraft have caused significant disruption to helicopter development programmes.

In 1974, a programme of research into future rotor blade design was launched. It was the British Experimental Rotor Programme (BERP) and its initial designs and their children have been in existence ever since. BERP I to BERP IV trace a sequence of rotor blade developments driven by improvements in technical ability and understanding, engineering materials and manufacturing processes.

While programmes such as BERP enable rotor blade designs to be refined and reset standards for forward flight, the helicopter is still dogged by the rotor aerodynamics to pursue high-speed flight: namely, the perennial advancing/retreating-blade problem and at high speeds the requirement on the main rotor for high values of disc tilt. Rotor disc tilt will cause the fuselage to follow and with significant nose-down attitudes the attendant fuselage attitude can contribute a negative lift thereby loading the main rotor even more. Consequently, in order for a pure helicopter to be effective with a high cruise speed, the payload carried must be sacrificed for the weight of the mechanical systems at an increasing rate, particularly if the cruise speed exceeds 170 knots. This difficulty has caused designers to examine different types of rotary-wing aircraft with emphasis on two, namely the Bell Boeing V22 tilt rotor and the Piasecki compound helicopter.

Now, in the early years of the twenty-first century, what can follow?

The pure helicopter is always required to have the capability to hover efficiently, but it is also required to fly efficiently at high forward speed and to be capable of achieving extremes of endurance and range. The phrase 'high forward speed' is rather vague but in pure helicopter terms it is relative to the normally accepted envelope of 170 knots. So the question is posed: is 170 knots the speed limit or are we looking at a faster transit speed? The follow-up to this is the influence this high speed will have on the operating cost and efficiency of the hover.

The operation of helicopters is progressively becoming more demanding on both the aircraft and the flight crew. The demands of operating from the confines of ships' flight decks, oil rigs or, as is becoming more prevalent, operating close to difficult terrain when visibility is poor, such as brownout, place a high requirement on sensors, so the increasing need to operate the helicopter close to the ground can avoid hazards such as cliff faces, hills, bridges and general operation in and out of cover.

It is unfortunate that the helicopter has attracted, in the past, bad publicity as regards safety of operation. This has not gone unnoticed and strenuous efforts are being made to make a helicopter as safe as possible. Indeed, airworthiness can formally prove that the helicopter is safe.

The gauntlet is thrown down to the pure helicopter to meet these challenges. There is no doubt that being able to construct new and improve existing theoretical methods will influence the design process with this knowledge and bring it to the aircraft.

The rotor blade has, pretty much so far, been a passive device. It retains its shape and only in recent years have the dynamic characteristics been used in the search for increased performance with aeroelastic tailoring. Progress now requires the consideration of active systems such as tip-jet blowing and boundary layer control. Also a considerable amount of effort is now being directed at trailing-edge morphing, where internal devices physically deform the aerofoil section. External flaps have been used for many years, but the ability to achieve these effects while avoiding an external device with its controls is seen as a step forward.

The rotor head is now increasingly using flexures rather than conventional hinges. These benefit the drag through aerodynamic cleanliness and remove a considerable burden of maintenance on the articulated rotor. A most influential benefit is the effect of a flexure system on the flight behaviour.

This naturally turns attention to the handling qualities of the helicopter. The control laws used to aid the pilot are now benefitting from the use of a greater level of embedded intelligence with the attendant ability to tailor the helicopter's flying qualities for a particular task.

A problem with the pure helicopter has always been the method of torque control to react against the torque supplied to the main rotor and to give the pilot sufficient yaw control – usually in difficult circumstances. The open tail rotor has been used for many years, but alongside this there has been the introduction of the fenestron, the NOTAR circulation-controlled tail boom, and maybe a return to the tip-jet-driven main rotor which has no need for a specialized yaw control device.

In addition to acceptable handling qualities, the aspect of safety should be considered. In the distant past, inspections and component life were used to determine flight safety. In more recent times, the ability of sensing systems to monitor the most critical of the helicopter components and to observe the vibration levels at different parts of the aircraft has created the Health and Usage Monitoring System (HUMS). This has improved the airworthiness levels considerably, but the future is now to use these systems to predict the degradation of a component or system so that replacement or servicing can be scheduled so that the aircraft can operate safely but the maintenance schedule will not be compromised. This is the prognostic version of HUMS. The observation of the aircraft by HUMS can provide the ability to measure fatigue giving the Fatigue and Usage Monitoring System (FUMS). All of this requires the use of the most modern and advanced electronic systems.

Or maybe new configurations will provide the way. A considerable number have been proposed or indeed flown over the years. The Sikorsky Advancing Blade Concept (ABC) of the late 1960s and early 1970s fitted to the S69 aircraft overcame the advancing/retreating-blade conflict by using a stiff coaxial rotor system which effectively relied on the advancing side of both rotors to provide the performance. Another direction was the return to the stopped rotor in the guise of the X Wing project of Sikorsky/NASA in the 1980s. It took off in the conventional way and then proceeded into forward flight under auxiliary propulsion. Use of blown blades allowed the rotor to achieve roll trim, especially as the rotor was slowed to a stop with the four-bladed rotor orientated in a 45° position giving rise to the X Wing name. The blowing sequences had to vary with blade radial position and blade azimuth making the computing effort considerable. The blades, which are normally kept in shape by the centrifugal force of rotation, are now required to support the aircraft by the structure only.

The tilt rotor concept returned and a great deal of effort was devoted to it in the 1970s, 1980s and 1990s, initially with the Bell XV-15 concept vehicle from which emerged the production version, which is the Bell Boeing V22 Osprey, now in service with the US Marine Corps – see Figure 9.1.

The tilt wing has not appeared in recent times. However, the one configuration which is seeing recent effort is the compound helicopter. The Piasecki Corporation has, for many years, worked on this type of vehicle, with particular attention paid to the tail unit providing the yaw control. By using a propeller with a cascade of vertical aerofoil louvres to deflect the downwash from the propeller in a sideways direction, the required yaw control is achieved. The louvres can be realigned so that the propeller can now act as a direct propulsion device removing this

Figure 9.1   Bell Boeing V22 Osprey

Figure 9.2   S-70 Blackhawk compound development aircraft (Courtesy Piasecki Corp.)

requirement from the main rotor. The most recent application is that of a modified S-70 Blackhawk as shown in Figure 9.2.

One significant benefit of a thrust compound helicopter is that by relieving the main rotor of the need to provide propulsion, the aircraft attitude can be kept under close control. Essentially the pilot now has the ability to control both components of force rather than the main rotor supplying the vector sum. This has immediate benefits for passengers in a civil aircraft and keeping weapons on target in a military helicopter. For both variants the ability to slow to a halt can now be achieved without the attendant problem of a high nose-up attitude which can severely reduce the pilot's view of the landing site.

The 1950s saw the development of the Fairey Rotodyne which is a compound helicopter but uses tip-jet drive in and around the hover. In forward flight it transfers to an autogyro mode whereby the power to the blade tip jets is cut off and the twin airscrews provide the necessary forward propulsion. It still holds interest – even today.

The most recent introduction to the compound helicopter stable is the Sikorsky X2 as shown in Figure 9.3.

This has a powered coaxial main rotor with a pusher propeller. On 15 September 2010 it achieved a speed of 262 knots.

Within days of the X2 achieving its high-speed record-breaking run, Eurocopter unveiled its X3 demonstrator – Figure 9.4.

This uses separate propellers mounted on stub wings either side of the main fuselage. It has a normal single main rotor. It is interesting that the idea of using forward-facing propellers to give yaw control and forward propulsion was also seen on the Fairey Gyrodyne.

A concept already mentioned is the stopped or slowed rotor. This can be seen in various developments, one being the CarterCopter which is based on the slowed rotor. A recent statement by Boeing had indicated a return to the idea of a stopped rotor. Boeing is developing a rotor that has a large thin disc at the centre of the main rotor hub. The blades are extendable/retractable being housed within the disc. If a stopped rotor is to work it will encounter extremely large advance ratios as the rotor slows to a halt. This places a high level of difficulty on the

Figure 9.3 Sikorsky X2 compound helicopter (Courtesy Ashish Bagai, Sikorsky Aircraft)

blades at 90° and 270° azimuth. The disc provides a safe location for the blades as the advance ratio increases beyond the normal limits.

One recent development has been the use of simulation on relatively inexpensive computer systems. It is, of course, imperative that the fidelity is of the highest quality. In the past this has required very expensive equipment; however, the quality of simulation software, which can be purchased on the high street, has improved a very great deal in the past few years and helicopter landings on ships are quite possible with inexpensive computing and software. The provision of a full moving base simulator will still be very expensive, but even that could fall to the emergence of computing power that would have been staggering only a few years ago.

The discussion has concentrated on the piloted helicopter. Rotorcraft with their unique ability to VTOL operation enable them to be considered for use as UAVs. An example is the Firescout as shown in Figure 9.5.

Figure 9.4 Eurocopter X3 demonstrator (Courtesy Eurocopter)

Figure 9.5   Northrop Grumman Firescout UAV helicopter (Courtesy US Navy)

It is designed to operate from ships, spending in excess of 8 hours on station. It can accept a range of payloads.

So far, the discussion has been based on the Earth. In the early years of the twenty-first century, the use of a rotorcraft for exploration of the planet Mars was considered. The weight limitations are of obvious major concern but the planet has a very different atmosphere. The 'air density' is about 1% of that on Earth. This immediately places a very high burden on the induced power in the hover, which it will have to do as a matter of course. It will be subjected to high wind speeds and so the VTOL capability will exact its price.

In the author's opinion, the helicopter will never replace the role of a civil fixed-wing aircraft, particularly long haul. The helicopter will supply a niche market and do it well and with increasing efficiency and safety. To ask for more is unrealistic. It will have to conform to evermore stringent regulations concerning pollution. This, of course, involves noise generation and many efforts have already been used to address this problem. An example is the scissor tail rotor of the Apache as shown in Figure 9.6.

Figure 9.6   Apache (scissor tail rotor)

This chapter is a discussion on what might happen. What will happen is another matter. Having worked in helicopters for 40 years now, I am very set in my ways. The baton now passes to another generation who will bring new and controversial ideas to the subject. If my 40 years have taught me anything, it is that helicopter engineers can be disruptive types who seem to specialize in asking the most awkward questions. Long may it continue!

References

1. Byham, G.M. (2003) The future of rotorcraft, *Aero. J.*, **1072**, 377–387.
2. Aiken, E.W., Ormiston, R.A., and Young, L.A. (2000) Future directions in rotorcraft technology at Ames research center, AHS 56th Annual Forum, Virginia Beach VA, May 2–4.
3. Young, L.A., Aiken, E.W., Gulick, V., Mancinelli, R., and Briggs, G.A. (2002) Rotorcraft as Mars scouts, IEEE Aerospace Conference, Big Sky, March 9–16.

Appendix

Performance and Mission Calculation[1]

A.1 Introduction

In the main body of the book, methods have been described which enable the calculation of the power necessary to allow a helicopter to operate at a given weight and at a given speed. The results of these calculations, when combined with appropriate engine data, enable the fuel consumption to be determined and a mission can then be 'flown' in a computer. The ability to perform these calculations now enables a project study to be carried out on a proposed helicopter design. The purpose of this appendix is to collate the various theories of the momentum/actuator disc into an overarching calculation scheme. Earlier chapters highlighted the fact that momentum theories are the simplest available and, in order to make the results more realistic, a scheme of factoring will be required. The simplicity of these methods makes them easy to implement but, in addition, prevents them from being used for calculations where the rotor is operating close to any limitations of the flight envelope. For normal operations, this limitation should not be necessary and the simplicity of these methods will permit parametric studies to be performed with speed and economy. The modern personal computer, and the software available, make these methods, described in this appendix, readily implementable. In this way, an overall picture of the proposed helicopter configuration and its ability to complete a given mission can be readily assessed.

The following sections describe the practical use of these methods of determining the power required and the consequent rate of fuel consumption for a helicopter of given weight and speed. The calculations use the momentum method and, because of this, should only be used for a general investigation of helicopter performance. They are unsuitable for investigating helicopter performance when the aircraft is approaching its flight envelope. The methods, as presented, are formulated for a single main and tail rotor configuration. However, because of their inherent simplicity, they may readily be adapted for other rotorcraft configurations such as the tandem.

The description of the various methods is arranged in separate sections. Each section deals with a particular aspect of the helicopter.

---

[1] Portions of this appendix have been taken from *The Foundations of Helicopter Flight*, by Simon Newman, Elsevier, 1994.

Basic Helicopter Aerodynamics, Third Edition. John Seddon and Simon Newman.
© 2011 John Wiley & Sons, Ltd. Published 2011 by John Wiley & Sons, Ltd.

A summary of the components of the calculation is:

1. Using the helicopter's drag and weight to determine the attitude of the main rotor disc and the thrust by balancing the force components.
2. Determination of the main rotor, induced, profile and parasite powers implementing the appropriate factors. These powers are summed to give the total power required to drive the main rotor.
3. The main rotor power is then converted to the equivalent torque which fixes the value of the tail rotor thrust necessary to trim the helicopter in yaw.
4. With the tail rotor thrust and forward speed now determined, again implementing appropriate factors, the induced and profile powers of the tail rotor can then be calculated (note: no parasite power).
5. The total helicopter power required can then be determined by summing the main and tail rotor powers together with that required to drive auxiliary services.
6. The losses in the transmission are then included as a multiplying factor which gives the power required of the engine(s).

A.2 Glossary of Terms

The nomenclature used in the following sections is presented in Tables A.1 and A.2.

A.3 Overall Aircraft

The initial calculation is of the helicopter drag which, together with the weight, will allow the thrust and forward tilt of the main rotor to be determined.

Table A.1 Rotor

	Main	Tail
Thrust	T_M	T_T
Blockage factor	B_M	B_T
Tip speed	V_{TM}	V_{TT}
Number of blades	N_M	N_T
Blade chord	C_M	C_{TL}
Rotor radius	R_M	R_T
Disc tilt	γ_s	
Thrust coefficient	C_{TM}	C_{TT}
Advance ratio	μ_M	μ_T
Advance ratio component parallel to disc	μ_{xM}	μ_{xT}
Advance ratio component perpendicular to disc	μ_{zM}	
Downwash	λ_{iM}	λ_{iT}
Profile drag coefficient	C_{D0M}	C_{D0T}
Induced-power factor	k_{iM}	k_{iT}
Induced power	P_{iM}	P_{iT}
Profile power	P_{PM}	P_{PT}
Parasite power	P_{PARAM}	
Total power	P_{TOTM}	P_{TOTT}

Table A.2   Overall helicopter

Drag @ 100 velocity units	D_{100}
Auxiliary power	P_{AUX}
Transmission loss factor	TRLF
Helicopter weight	W
Helicopter drag	D

Drag can be specified in several ways; however, the method described here uses the drag force at a reference speed of 100 units (D_{100}) at ISA sea-level air density as a basis. (This calculation can be readily adapted to any other specification of drag.) The drag of the helicopter is then calculated by factoring the D_{100} value with the square of the forward speed and linearly with respect to the air density, as follows:

$$D = D_{100} \cdot \left(\frac{V}{100}\right)^2 \cdot \sigma \tag{A.1}$$

A.3.1   Main Rotor

A.3.1.1   Calculate Main Rotor Thrust and Disc Attitude

The force balance diagram for the main rotor is shown in Figure A.1.
   Resolving vertically:

$$T\cos(\gamma_S) = W \tag{A.2}$$

Resolving horizontally:

$$T\sin(\gamma_S) = D \tag{A.3}$$

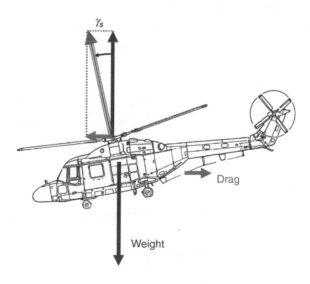

Figure A.1   Calculation of main rotor disc tilt and thrust

From these we obtain for the disc tilt:

$$\gamma_S = \tan^{-1}\left(\frac{D}{W}\right) \tag{A.4}$$

The main rotor thrust, with the application of blockage, becomes:

$$T = B_M \cdot \sqrt{W^2 + D^2} \tag{A.5}$$

Rotor blockage represents the download on the fuselage due to the rotor downwash and is applied to the main rotor thrust as a multiplying factor using the B_M factor.

The induced velocity of the main rotor can now be determined. Since actuator disc theory is being used, the advance ratio components parallel to and normal to the main rotor disc plane are needed together with the thrust coefficient.

The advance ratio is defined by:

$$\mu_M = \frac{V}{V_{TM}} \tag{A.6}$$

Resolving parallel to the rotor disc:

$$\mu_{XM} = \mu_M \cos(\gamma_S) \tag{A.7}$$

Resolving perpendicular to the rotor:

$$\mu_{ZM} = \mu_M \sin(\gamma_S) \tag{A.8}$$

A.3.1.2   Calculate Main Rotor Downwash

With the advance ratio components evaluated, the main rotor downwash can now be calculated using the iterative technique described in an earlier chapter:

$$\lambda_{iM\ NEW} = \lambda_{iM\ OLD} - \frac{\left[\lambda_{iM\ OLD} - \dfrac{C_T}{4\sqrt{\mu_{XM}^2 + (\lambda_{ZM} + \lambda_{iM\ OLD})^2}}\right]}{1 + \dfrac{(\mu_{ZM} + \lambda_{iM\ OLD})C_T}{4\left(\sqrt{\mu_{XM}^2 + (\lambda_{ZM} + \lambda_{iM\ OLD})^2}\right)^3}} \tag{A.9}$$

A good starting value, for the iteration, is that for hover, namely:

$$\lambda_{i\ OLD} = \frac{1}{2}\sqrt{C_T} \tag{A.10}$$

This now determines the downwash λ_{iM}.

A.3.1.3   Assemble Main Rotor Powers

The main rotor power components can now be calculated.
Induced (*note that the induced power factor k_{iM} is included*):

$$P_{iM} = k_{iM} \cdot T_M \cdot V_{TM} \cdot \lambda_{iM} \tag{A.11}$$

Profile:

$$P_{PM} = \frac{1}{8}\rho V_{TM}^3 \cdot N_M C_M R_M \cdot C_{D0M}\left(1 + 4.7\mu_{XM}^2\right) \tag{A.12}$$

Parasite:

$$P_{PARAM} = D \cdot V \tag{A.13}$$

Summing for the total:

$$P_{TOT\,M} = P_{iM} + P_{PM} + P_{PARAM} \tag{A.14}$$

A.3.2   Tail Rotor

The main rotor torque is now obtained from the total main rotor power, from which the tail rotor thrust value necessary for torque balance is calculated.

A.3.2.1   Calculate Tail Rotor Thrust

$$T_T = B_T\left(\frac{P_{TOTM}}{\Omega_M \cdot l_{BOOM}}\right) \tag{A.15}$$

(Note the inclusion of the blockage factor B_T for the tail rotor.)
   The determination of the induced velocity of the tail rotor is the next part of the calculation. The advanced ratio components together with the thrust coefficient of the tail rotor are now required:

$$\mu_{XT} = \frac{V}{V_{TT}} \tag{A.16}$$

$$\mu_{ZT} = 0 \tag{A.17}$$

   The tail rotor is only required to develop a thrust, normal to the helicopter's centreline, so has no need for any disc tilt; in consequence the disc plane is assumed parallel to the flight path.

A.3.2.2   Calculate Tail Rotor Downwash

Using the same iterative method, the tail rotor downwash, λ_{iT}, is then calculated.

A.3.2.3   Assemble Tail Rotor Powers

The tail rotor power components are now calculated.
  Induced:

$$P_{iT} = k_{iT} \cdot T_T \cdot V_{TT} \cdot \lambda_{iT} \tag{A.18}$$

Profile:

$$P_{PT} = \frac{1}{8} \rho V_{TT}^3 \cdot N_T C_{TL} R_T \cdot C_{D0T} \left(1 + 4.7\mu_{XT}^2\right) \tag{A.19}$$

Total:

$$P_{TOTT} = P_{iT} + P_{PT} \tag{A.20}$$

(Note that there is no parasite power for the tail rotor as the main rotor is assumed to be responsible for overcoming the parasite drag of the aircraft.)

A.3.3   Complete Aircraft

The total power required is now – calculated by summing the total powers for the main and tail rotors together, to which is added the power necessary to drive any auxiliary services (P_{AUX}) such as oil pumps and electrical generators. It is to be expected that some losses will occur in the transmission and, to account for this, a factor is applied giving the power required from the engines as follows.

A.3.3.1   Assemble Overall Aircraft Powers and Allow for Transmission Losses

$$P_{TOTAL} = \text{TRLF} \times \left(P_{TOTM} + P_{TOTT} + P_{AUX}\right) \tag{A.21}$$

A.3.4   Example of Parameter Values

Table A.3 suggests values of the various factors applied – which are used in the example mission calculation.

  As already described, rotor blockage is a result of the downwash impinging on the fuselage, for the main rotor; however, there is also a blockage factor for the tail rotor to account for its interaction with the fin. The effect of blockage is to generate a force on the fuselage/fin in

Table A.3

Parameter	Symbol	Value
Rotor blockage	B_M	1.05
	B_T	1.1
Induced-power factor	k_{iM}	1.1
	k_{iT}	1.2
Profile drag coefficient	C_{D0M}	0.011
	C_{D0T}	0.012

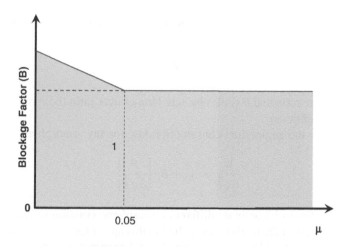

Figure A.2 Variation of blockage factor

opposition to the respective rotor thrust directions. Therefore, in order to achieve the required component of force from either rotor, the thrust must exceed this by an amount equal to the download on the fuselage or fin. As the effect of these downwashes on the rotors will be altered by the superimposing of the forward flight velocity, the blockage effect will diminish with an increase in this forward flight velocity. In essence, the rotor downwash is carried increasingly downstream of the fuselage/fin, eventually causing no real fuselage/fin download from the main/tail rotor wake.

The application of this rotor blockage is to use a factor which multiplies the desired net thrust to account for the loss. As the forward velocity of the helicopter increases, the blockage factor will reduce from its specified hover value to unity. The variation is specified in a simple manner and for this example is based on the respective advance ratio. The blockage value refers to the hover condition, where the interaction with the fuselage/fin is greatest, and linearly decreases to unity at an advance ratio of 0.05, remaining at unity for higher advance ratios. This is illustrated in Figure A.2.

A.4 Calculation of Engine Fuel Consumption

At this point of the calculation the total power required of the engine(s), for the given weight and flight condition, is known. We now use this information to determine the fuel consumption.

In most instances, engine fuel consumption data are given in terms of the specific fuel consumption (sfc) (kg/h/kW) for a corresponding power setting (P). A small adjustment of the data permits a simple method to be used to calculate the fuel consumption of a gas-turbine engine from the specified amount of power.

The concept of fuel flow of an engine (W_f) (kg/h) is now introduced. It is readily seen that it is obtained from the product of the sfc and the power. By plotting the resulting fuel flow against power, a variation very close to linear can be seen and, hence, can be specified by a straight line equation which can be determined by linear regression (least squares). The resulting linear variation makes the fuel consumption calculation very straightforward.

To make the method even more useful, the operating altitude and temperature will need to be incorporated into the calculation. If the fuel flow v power variation is plotted for each atmospheric condition, a series of straight line fits will result. However, these straight line fits will collapse close to one single straight line if the fuel flow and engine power are normalized by the factor:

$$\delta\sqrt{\theta} \tag{A.22}$$

where δ is the pressure ratio and θ is the absolute temperature ratio (both relative to ISA sea-level atmosphere conditions).

We can thus define the engine fuel consumption law for any atmospheric condition as:

$$\frac{W_f}{\delta\sqrt{\theta}} = A_E + B_E\left[\frac{P}{\delta\sqrt{\theta}}\right] \tag{A.23}$$

This gives the ability to incorporate different atmospheric conditions into the calculation method. As with the rest of the methods described in this appendix, its simplicity requires that for this fuel consumption calculation, it is assumed that the engine(s) is(are) not operating close to a limit.

The resulting straight line fit has a positive intercept on the fuel-flow axis, defined by the term A_E. This has an important influence on the optimization of fuel consumption for a multi-engined helicopter. If we have a helicopter which has N engines, each combining to give a total power production of P, then it follows that each engine must generate a power of P/N whereupon the total fuel consumption for all N engines combined is given by:

$$\frac{W_f}{\delta\sqrt{\theta}} = N\left\{A_E + B_E\left[\frac{P}{N}\cdot\frac{1}{\delta\sqrt{\theta}}\right]\right\} \tag{A.24}$$

that is:

$$W_f = N\cdot A_E\cdot\delta\sqrt{\theta} + B_E\cdot P \tag{A.25}$$

The result of the first term on the right-hand side of (A.24) means that, from (A.25), it can be seen that for a given power requirement, the smaller the number of engines, the lower the fuel consumption. In consequence, as a helicopter design develops, if the optimizing for fuel consumption is paramount, a minimum number of engines capable of providing sufficient power should be the choice. This may be in direct conflict with the other requirements, particularly with the performance of the helicopter having sustained an engine failure when the design will tend to move in the opposite direction – that is, to have a maximum number of engines. As can be seen, the selection of the engine provision for a multi-engine helicopter is therefore not so simple.

A.5   Engine Limits

So far, as regards the engine installation, the performance has been focused on the fuel consumption. During its operational life, a gas-turbine engine might be required to operate outside of normal continuous limits. In such circumstances, it will have limitations placed on it which are determined by the permissible operating temperature of the turbine section. So if the engine is required to operate at a power above the normal continuous limit, providing this occurs for a specified limited time period, it is possible to achieve this without causing permanent damage. It may happen that a situation arises which can be considered to be an emergency and in order to save the helicopter, excessive wear or indeed damage to the engine(s) may be the only possible choice. These are also catered for, but the time limits are necessarily short. To illustrate this, typical examples of such power limitations are as follows.

A.5.1   Maximum Continuous Power Rating

This is the maximum power at which an engine can operate continuously. Consequently, it does not have a time constraint.

A.5.2   Take-Off or 1 Hour Power Rating

This rating is applicable for the higher power situations such as operation at high altitude and/or ambient temperature and particularly for take-off and hover. Time limits of approximately 1 hour (sometimes $1/2$ hour) are allowed before the engine must revert to a lower power setting. *(A working figure is 10% above the maximum continuous rating.)*

A.5.3   Maximum Contingency or $2^1/_2$ Minute Power Rating

By its title, this power rating is used in contingency situations, such as the loss of an engine. The time limit is considerably shorter and usually for a period of 2 to 3 minutes. *(A working figure is 20% above the maximum continuous rating.)* Because of the high level of power increase it is quite possible that an engine inspection be considered.

A.5.4   Emergency or $1/_2$ Minute Power Rating

This is a rating used only as a last resort, when saving the helicopter is the priority. Engine damage is a real possibility for this situation. The time limit is very short (30 seconds) since engine failure is a real consideration. *(A working figure is 30% above the maximum continuous rating.)*

To illustrate the need for such an excess of power the following situation is provided as an example. Consider a twin-engine naval helicopter which suffers an engine loss in a condition requiring high power – hovering at a high all-up weight, for instance. If this occurs over the sea then the pilot may be forced to lower the helicopter onto the sea surface. To achieve a take-off from the sea with an engine lost will require a reduction in all-up weight, so jettisoning as much weight as possible will be necessary. Emergency power will be required for a take-off on a single engine from the water. After retrieving the helicopter from such a dire situation and returning to base/ship, the engine(s) will probably require extensive maintenance and refurbishment. The damage may be such that it is beyond repair and will need to be scrapped.

A.6   Calculation of the Performance of a Helicopter

To illustrate the method, the following calculations, based on a small utility helicopter, are presented.

The helicopter data used are given in Table A.4.

The fuel-flow variation with power, presented in (A.25), was obtained from public domain information (a leaflet obtained at an air show) and, using linear regression, the following

Table A.4

Parameter	Symbol	Value
Rotor data:		
Number of blades	N_M	4
	N_T	4
Chord (m)	C_M	0.394
	C_{TL}	0.180
Radius (m)	R_M	6.4
	R_T	1.105
Tip speed (m/s)	V_{TM}	218.69
	V_{TT}	218.69
Blockage	B_M	1.05
	B_T	1.1
Induced-power factor	k_{iM}	1.1
	k_{iT}	1.2
Profile drag coefficient	C_{D0M}	0.011
	C_{D0T}	0.012
Fuselage data:		
Tail boom length (m)	l_{BOOM}	7.66
D_{100} (N)	D_{100}	6226.9
Auxiliary power (kW)	P_{AUX}	26.1
Transmission loss factor	TRLF	1.04
Engine data:		
Number of engines	N_E	2
Intercept	A_E	46.5
Slope	B_E	0.24

equation was obtained:

$$\frac{W_f}{\delta\sqrt{\theta}} = 46.5 + 0.24\,\frac{P}{\delta\sqrt{\theta}} \tag{A.26}$$

Using the data detailed above, the variation of the main rotor power components with forward speed is shown in Figure A.3, namely induced, profile, parasite and total.

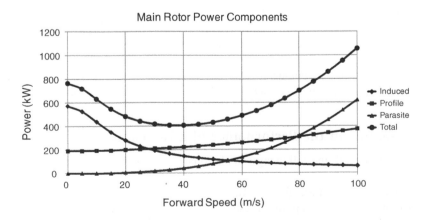

Figure A.3 Main rotor power components

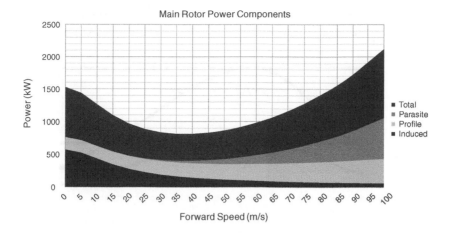

Figure A.4 Main rotor power components – cumulative

If these power components are viewed cumulatively, as in Figure A.4, the build-up of the total power can be seen.

Having determined the total main rotor power, the total tail rotor power, the auxiliary power and the influence of the transmission losses, they can be combined to give the total overall power required of the engine(s).

This total power variation of the complete helicopter with forward speed is shown in Figure A.5.

This power distribution now enables the fuel consumption to be calculated and by selecting a given weight of fuel to be consumed, the helicopter endurance and range can be calculated for the range of speeds. *(For these calculations, the helicopter weight is considered constant.)* Endurance, by definition, is the time required to consume that specified amount of fuel, and range is defined as the distance covered while consuming that fuel amount. It is apparent that endurance is focused on minimizing the rate of fuel usage with time, while range includes both time and speed and is therefore a compromise. (A fuel usage of 100 kg is assumed for these calculations.) The endurance is shown in Figure A.6, and the corresponding range (km) is

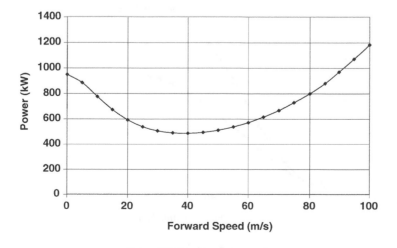

Figure A.5 Total power consumption

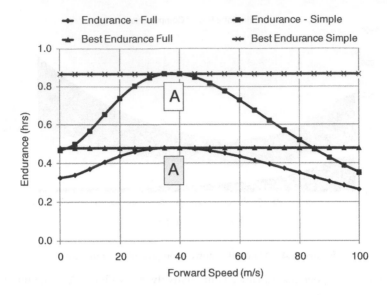

Figure A.6   Maximum endurance

shown in Figure A.7. Each figure contains two plots referring to either the full fuel-flow law as defined in Equation A.26, or a modification to that law where the intercept (A_E, or fuel flow at zero power) is set to zero. (Inspection shows that this corresponds to a constant sfc.) The lower curve corresponds to the full law ($A_E \neq 0$) and the upper curve to the modified law ($A_E = 0$) – where the fuel consumption rate (fuel flow) is smaller.

The positions of maximum endurance and range are indicated in the figures by a letter. Figure A.6 shows that the change in the fuel-flow law does not alter the best endurance speed of

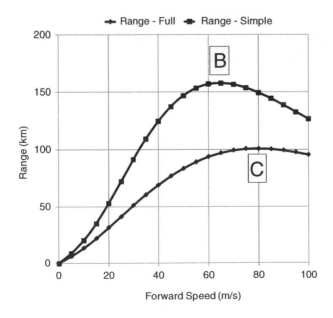

Figure A.7   Maximum range

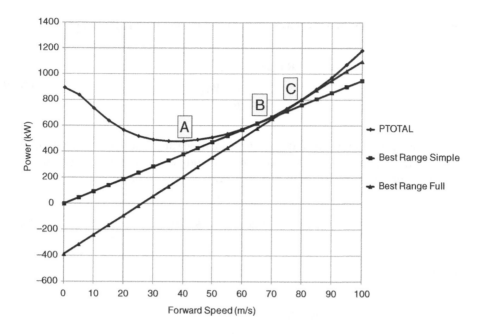

Figure A.8   Geometrical construction of maximum range speed

38 m/s (A), which, not surprisingly, corresponds to minimum power. However, because of the different fuel consumption rates, the change does have an influence on the best-range speed increasing it from 65 m/s (B) to 80 m/s (C) albeit with a smaller range.

Figure A.8 shows the best-range speeds, also indicated by B, and C, on the total power v forward speed curve.

In fact, the two points, B and C, corresponding to these 'best'-range conditions, can be determined via a simple geometrical construction as the analysis below shows. Using the definitions in Table A.5, we have the following.

For point A (maximum endurance), the definition for sfc is:

$$S = \frac{W_{\text{FUEL}}}{P \cdot \text{time}} \tag{A.27}$$

that is:

$$\text{Endurance} = \frac{W_{\text{FUEL}}}{S} \cdot \frac{1}{P} \tag{A.28}$$

Since W_{FUEL} and S are fixed values, the endurance is a maximum when P is minimum, that is point A is the minimum point on the power v velocity curve.

Table A.5

W_{FUEL}	Fuel weight (fixed)
P	Power
V	Forward speed
E	Endurance
R	Range

For point B (maximum range – constant sfc, $A_E = 0$), the range is given by:

$$\text{Range} = \text{time} \cdot V$$
$$= \frac{V}{P} \cdot \frac{W_{\text{FUEL}}}{S} \tag{A.29}$$

that is the range is a maximum (since W_{FUEL} and S are fixed values) when P/V is a minimum, so point B is located where the tangent, drawn from the origin, touches the power curve.

For point C (maximum range – full fuel-flow law), in this we have:

$$W_{\text{FUEL}} = \text{time} \cdot (A_E + B_E \cdot P) \tag{A.30}$$

Therefore:

$$\text{Range} = \text{time} \cdot V$$
$$= \frac{V}{(A_E + B_E \cdot P)} W_{\text{FUEL}} \tag{A.31}$$
$$= \frac{W_{\text{FUEL}}}{B_E} \cdot \frac{V}{\left(\dfrac{A_E}{B_E} + P\right)}$$

This result is very similar to that defining point B except there is an additional term in the denominator (A_E/B_E). This means that the tangent has to be drawn from the point $(0, -A_E/B_E)$. These constructions for the points B and C are shown in Figure A.8.

A.6.1   Influence of Wind

If the helicopter is flying into a headwind of V_W, the above formula for range becomes:

$$\text{Range} = \text{time} \cdot (V - V_W)$$
$$= \frac{(V - V_W)}{(A_E + B_E \cdot P)} \cdot W_{\text{FUEL}} \tag{A.32}$$
$$= \frac{W_{\text{FUEL}}}{B_E} \cdot \frac{(V - V_W)}{\left(\dfrac{A_E}{B_E} + P\right)}$$

This means that the tangent should be drawn from the point $(V_W, -A_E/B_E)$, as shown in Figure A.9.

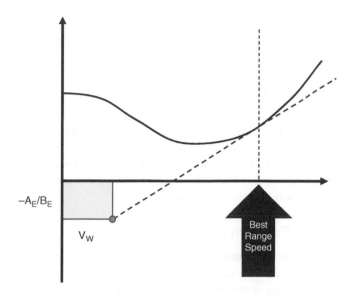

$-A_E/B_E$

V_W

Best Range Speed

Figure A.9 Best-range speed with wind

A.7 Mission Analysis

In order to examine the ability of a helicopter design to perform a given mission, the calculations described – where engine power and fuel consumption can be determined at any flight speed – can be assembled so the helicopter model can 'fly' a mission in a computer. The ease and immediacy of this procedure make it of direct use to a project assessment. There is one important consideration to be made. As the helicopter consumes fuel, the weight changes, which in turn affects the fuel consumption itself. So, in an ideal world, the calculation becomes circular. However, if the mission is 'flown' in small time steps, or mission legs, then an iterative scheme can be used to obtain an estimate of the fuel consumption rate over the leg/time period. The mission is then assembled by linking these individual mission legs where the value of the helicopter weight at the end of a particular leg becomes the start value for the succeeding leg. (Each leg will be defined by a fixed flight condition, particularly forward speed. However, if a climb or descent – at constant speed - is required the iterative scheme can still be used.) The mission may contain discrete weight changes of payload, such as changes in passenger/cargo payload or the deployment of ordnance. These can be incorporated by placing any such occurrence at the join of two mission legs and the weight change made in moving from one to its successor.

The iterative scheme to account for the changing helicopter weight is now described. For each leg the procedure begins with the calculation of the power and fuel consumption at the start weight of the leg. The duration of the leg can be obtained either by the time being explicitly stated or by dividing the range by the speed to give the time. This enables a first estimate of the weight change over the leg to be calculated. By subtracting half of that fuel usage from the start weight, a revised helicopter weight is thus obtained. Taking this new value of aircraft weight the calculation process is repeated and a new estimate for fuel usage is thus obtained. The two values of the fuel usage, over the leg, at the two weights are then compared. If they differ within a specified tolerance, the process is seen to have converged and the final estimate is adopted. If the fuel usage values do not lie within the tolerance, a revised mean aircraft weight is adopted (by subtracting half of the latest fuel usage value from the start weight of the particular mission leg) and the process is repeated. This iteration continues until convergence to within the required tolerance is achieved.

A.7.1   Calculation Method

The calculation procedure is now presented, in schematic form, and is the basis for a flow diagram. Each part of the calculation is presented as a complete entity, in programming terms a function procedure or subroutine, with the appropriate input and output data being specified. This gives a segmented structure to the overall method, which is recommended for implementation in a computer program. The main program forms the input/output of data leaving separate routines to perform the detailed calculations.

A.7.2   Atmospheric Parameters

This determines the air parameters used by the calculations to be determined directly from the altitude and ambient temperature. An ISA atmospheric model is used. The helicopter is assumed to remain within the troposphere. Normally this is the case but in some instances helicopters have achieved very high altitudes; however these are truly exceptional and do not invalidate the model proposed.

A.7.2.1   Input

- Altitude
- Sea-level air temperature (absolute temperature at sea level)
- Sea-level air density.

A.7.2.2   Output

- Density ratio (σ)
- Absolute temperature ratio (θ)
- Pressure ratio (δ).

A.7.2.3   Calculation

The absolute temperature ratio is given by:

$$\theta = \frac{\text{AbTmp}_{\text{Sea Level}} - \text{Altitude} \times \text{LapseRate}}{\text{AbTmp}_{\text{Sea Level}}}$$
$$= 1 - \frac{\text{Altitude} \times \text{LapseRate}}{\text{AbTmp}_{\text{Sea Level}}} \tag{A.33}$$

The pressure ratio by:

$$\delta = \theta^{5.256} \tag{A.34}$$

The relative density by:

$$\sigma = \theta^{4.256} \tag{A.35}$$

The lapse rate of 6.5 °C per kilometre is used in the calculations.

A.7.3   Downwash Calculation

This is the application of momentum theory to the downwash calculation.

A.7.3.1   Input

- Thrust coefficient (C_T),
- Advance ratio components:
  - parallel to the rotor disc (μ_x)
  - perpendicular to the rotor disc (μ_z).

A.7.3.2   Output

- Downwash (λ_i).

A.7.3.3   Calculation

Start value (hover):

$$\lambda_{iOLD} = \frac{1}{2}\sqrt{C_T} \tag{A.36}$$

†

$$\Delta\lambda_i = \left[\frac{\lambda_{i\,OLD} - \dfrac{C_T}{4\sqrt{\mu_X^2 + (\mu_X + \lambda_{i\,OLD})^2}}}{1 + \dfrac{(\mu_Z + \lambda_{i\,OLD})C_T}{4\sqrt{\mu_X^2 + (\mu_Z + \lambda_{i\,OLD})^2}^{-3}}}\right] \tag{A.37}$$

$$\lambda_{i\,NEW} = \lambda_{i\,OLD} - \Delta\lambda_i \tag{A.38}$$

Has the iteration reached convergence?

$$|\lambda_{i\,NEW} - \lambda_{i\,OLD}| < \lambda_{Tol} \tag{A.39}$$

(Note the application of the absolute value of the difference.)
If NO then reset the downwash value and calculate the next estimate:

$$\lambda_{i\,OLD} = \lambda_{i\,NEW} \tag{A.40}$$

GO TO †
If YES then convergence has been achieved:

$$\lambda_i = \lambda_{i\,NEW} \tag{A.41}$$

EXIT

A.8   Helicopter Power

This is the central helicopter power calculation.

A.8.1.1   Input

- Aircraft all-up weight
- Forward speed
- Atmospheric data
- Aircraft data.

Suffix M refers to the main rotor, while T refers to the tail rotor.

A.8.1.2   Output

- Helicopter power.

A.8.1.3   Calculation

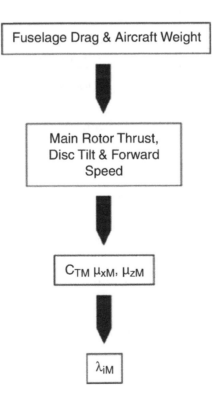

$$P_{iM}$$
$$+ P_{PM}$$
$$+ P_{PARAM}$$
$$\boxed{P_{TOTM}}$$

(A.42)

$$
\begin{aligned}
T_T &= \frac{Q_M}{l_{BOOM}} \\
&= \frac{P_{TOTM}}{\Omega_M \cdot l_{BOOM}} \\
&= \frac{P_{TOTM} \cdot R_M}{V_{TM} \cdot l_{BOOM}}
\end{aligned}
$$

(A.43)

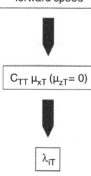

Tail Rotor Thrust (assume no disc tilt) & forward speed

$C_{TT} \, \mu_{xT} \, (\mu_{zT} = 0)$

λ_{iT}

$$P_{iT}$$
$$+ P_{PT}$$
$$\boxed{P_{TOTT}}$$

(A.44)

$$P_{TOTM}$$
$$+ P_{TOTT}$$
$$+ P_{AUX}$$
$$\boxed{P_{TOTAL}}$$

(A.45)

$$P_{REQ} = TRLF \cdot P_{TOTAL}$$

(A.46)

EXIT

A.9   Fuel Flow

This calculates the fuel consumption rate from the power requirements using the engine laws.

A.9.1.1   Input

- Engine power (P)
- Number of engines (N)
- Atmospheric pressure ratio (δ)
- Atmospheric temperature ratio (θ)
- Engine performance coefficients (A_E, B_E).

A.9.1.2   Output

- Fuel flow for the stated power required (W_f).

A.9.1.3   Calculation

$$W_f = \delta\sqrt{\theta} \cdot N_E \cdot A_E + P_{Req} \cdot B_E \tag{A.47}$$

EXIT

A.10   Mission Leg

This calculates the fuel usage over a mission component, or leg.

A.10.1.1   Input

- Start and finish altitudes (power and fuel flow are averaged between these two conditions)
- Start all-up weight (AUW$_{START}$)
- Forward speed (V)
- Time or distance of leg (e.g. 5 min hover or 20 km cruise at 100 m/s).

A.10.1.2   Output

- Fuel used during mission leg.

A.10.1.3 Calculation

Set variables to the conditions at the start of the leg.

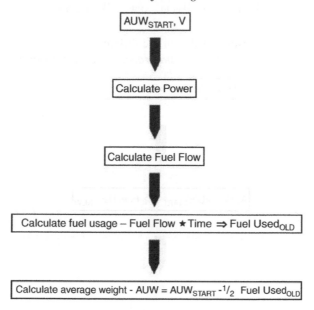

$\dagger$

Calculate the fuel usage with the revised helicopter weight.

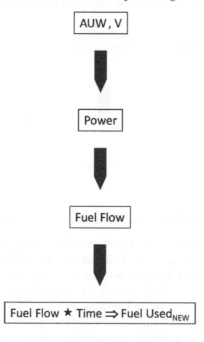

Has the fuel usage estimate converged?

$$|W_{\text{FUEL}-\text{OLD}} - W_{\text{FUEL}-\text{NEW}} < W_{\text{FUEL}-\text{Tol}} \qquad (A.48)$$

(Note the application of the absolute value of the difference.)

If NO, then reset the helicopter weight and proceed to a new estimate.

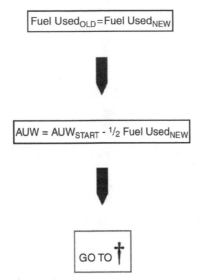

If YES, then fuel usage calculation has converged and therefore the weight change over the leg can be determined.

AUW$_{\text{FINISH}}$ = AUW$_{\text{START}}$ - Fuel Used$_{\text{NEW}}$

EXIT

A.11   Examples of Mission Calculations

In order to demonstrate the calculation procedure, when applied to a mission, an example is presented here. It comprises a helicopter of similar size to the Westland WG13 aircraft (from which the Lynx was developed) flying a search, rescue and recovery mission.

It is purely fictitious and does not represent any particular existing mission. It is used to demonstrate how the calculation procedure can be used to assess the performance of a helicopter engaged on a particular mission. While the mission calculation was performed using the Westland WG13/Lynx as the datum aircraft, the helicopter was modified in four ways, each one illustrating a parametric change which a designer could select. The results show the effect of such changes on the fuel usage of the helicopter. The ability to make rapid changes to a

Table A.6

Case no.	1	2	3	4	5
D_{100}	6227	12454	6227	6227	6227
Main rotor radius	6.401	6.401	6.901	6.401	6.401
Tail rotor radius	1.105	1.105	1.605	1.105	1.105
Tail boom length	7.66	7.66	8.66	7.66	7.66
Number of engines	2	2	2	1	3

helicopter design, and see the result, provides feedback as to the merits of any such change or, by scrutinising the various outputs of the calculation, the reasons why a change has no useful effect.

1. This is the basic aircraft.
2. This has the aircraft parasitic drag doubled via the D_{100} term.
3. This has the main and tail rotor radii increased by 0.5 m, with the consequent increase in tail boom length of 1 m.
4. The number of engines is reduced to 1. *No allowance has been made for the ability of the single engine to generate sufficient power. Only the fuel consumption has been studied.*
5. The number of engines is increased to 3. *Again, no allowance has been made for the ability of the single engine to generate sufficient power. Only the fuel consumption has been studied.*

The five sets of parameter changes are given in Table A.6.

A.12　Westland Lynx – Search and Rescue

A.12.1　Description of the Mission

The mission is divided into eight distinct legs which are given in Table A.7.

Table A.7

Leg	Phase	Altitude	Speed	Time	Distance	Remarks
1	Take-off	Sea level	Hover	5 min		
2	Flight out	Sea level	70 m/s		100 km	
3	Loiter	Sea level	50 m/s	5 min		
4	Sustained hover	Sea level	Hover	10 min		Drop medic (lose 80 kg)
5	Loiter	Sea level	50 m/s	10 min		
6	Sustained hover	Sea level	Hover	10 min		Retrieve medic and patient (acquire 160 kg)
7	Flight back	Sea level	70 m/s		100 km	
8	Land	Sea level	Hover	5 min		

A.12.2   Fuel Consumption

The summary of the fuel consumption for all eight mission legs and for each of the five helicopter configurations is tabulated in Table A.8. The total fuel consumption is included, expressed as a mass and as a percentage relative to the basic helicopter (case 1).

As can be seen in case 2, changes in D_{100} increase the fuel consumption by 12% which is consistent with the amount of time spent at high speed where parasite power dominates. Changes to rotor size, in case 3, produce virtually no change in fuel usage. This is explained by the fact that an increase in disc area reduces the hover power (induced component), but the consequent increase in blade area (increased rotor radius and identical blade chord) increases the profile power, particularly at the high speeds, and for this mission the two influences effectively cancel out. The changes in the number of engines, cases 4 and 5, are the most influential, showing the advantage, from a fuel consumption viewpoint, of operating on a single engine, and, on the same basis, the disadvantage of carrying a third engine.

Table A.8

Fuel	1	2	3	4	5
1	26.7	26.7	25.8	22.9	30.6
2	99.3	121.3	100.7	80.9	117.8
3	17.6	19.3	17.8	13.8	21.5
4	52.2	51.9	50.4	44.7	59.7
5	35.1	38.4	35.4	27.5	42.8
6	50.6	50.3	49.0	43.2	57.9
7	98.6	120.4	100.1	80.3	116.9
8	25.2	25.0	24.4	21.6	28.7
TOTAL	405.4	453.3	403.6	334.9	475.9
% of 1	100.0%	111.8%	99.6%	82.6%	117.4%

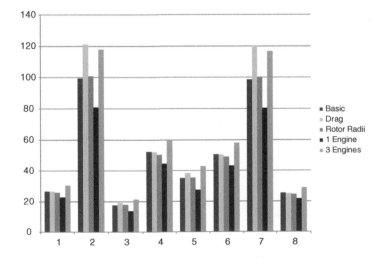

Figure A.10   Fuel usage summary

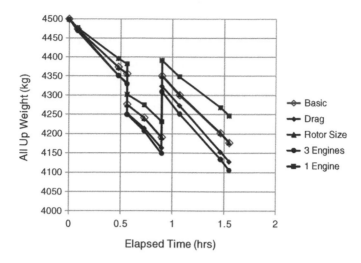

Figure A.11   All-up weight variation

These results are plotted in Figures A.10 and A.11. Figure A.10 shows the fuel usage for the eight legs and, under each leg, the five helicopter configurations ordered left to right. Figure A.11 shows the helicopter weight variation with time for each configuration over the complete mission.

Index

Basic Helicopter Aerodynamics, Third Edition. John Seddon and Simon Newman.
© 2011 John Wiley & Sons, Ltd. Published 2011 by John Wiley & Sons, Ltd.

Printed and bound by CPI Group (UK) Ltd, Croydon, CR0 4YY

27/10/2024

14580199-0001